Telecommunication: New Dynamics and Driving Forces

Edited by

Annaflavia Bianchi
CURDS, University of Newcastle, Newcastle upon Tyne, UK

Giuseppe Richeri
Centro Studi San Salvador, Telecom Italia, Venice, Italy
Università di Firenze, Florence, Italy

1996

IOS Press
Ohmsha

Amsterdam, Oxford, Tokyo, Washington, DC

ISBN 90 5199 256 4 (IOS Press)
ISBN 4 274 90085 1 C3000 (Ohmsha)
Library of Congress Catalogue Card Number 96-75083

Publisher
IOS Press
Van Diemenstraat 94
1013 CN Amsterdam
Netherlands

Distributor in the UK and Ireland
IOS Press/Lavis Marketing
73 Lime Walk
Headington
Oxford OX3 7AD
England

Distributor in the USA and Canada
IOS Press, Inc.
P.O. Box 10558
Burke, VA 22009-0558
USA

Distributor in Japan
Ohmsha, Ltd.
3-1 Kanda Nishiki - Cho
Chiyoda - Ku
Tokyo 101
Japan

PRINTED IN THE NETHERLANDS

umail

L

Telecommunications Policy Research

ISSN: 1385-1527

Preface and acknowledgements

This volume presents the results of research carried out by a selected group of scholars from different branches of the rich and rapidly developing field of telecommunications. It is an up-to-date dialogue between resaerchers and analysts on policy-relevant telecommunications issues and public and private sector decision-makers engaged in making telecommunications policy.

We are grateful to the editorial committee which provided essential stimulus to the research and enriched the debate that followed.

We are also extremely grateful to Dr. Derek Manson-Smith, who carried out the technical editing of this volume with remarkable patience and proficiency.

The Istituto di Economia dei Media, Milan, generously supported the research work.

Annaflavia Bianchi
Giuseppe Richeri

February 1996

Contents

Introduction

For this volume, we selected eight contributions to themes in the current debate on telecommunications. They focus on the two major issues of *sectoral convergence* – looking at the new challenges to regulation in Europe and at company strategies for alliance and competition at the national and global level – and *competition* – discussing interconnection of both local/long distance, incumbent/new entrant, and the definition of universal service in a liberalised framework, and also the new dynamics between network and service competition, with special attention on the role played by national telecommunications operators.

Each theme concerns both corporate strategy in the international environment and public policy responses to shifting realities.

The contributions present well-argued considerations that outline and provide evidence of the current phase of transformation which the sector entered in the mid-80s. This is particularly true for the complex European political and institutional system. In most of the European countries, the transformation process was stimulated by the action of the European Commission. Its directives and research and pilot-project support programmes provided guidance and stimulus for a new institutional and organisational structure for national telecommunications systems. It is well-known that the characteristic element of this process is the opening of the national monopolies to competition, in services and infrastructures.

The main interlocutors of the Commission were the national governments of the member states – each of which had strong relationships with telecommunications providers and manufacturers. The relationships between the Commission and individual member states developed with different degrees of intensity taking, case by case, the form of conflict, stimulus and cooperation, and passive adaptation. Three main typologies of behaviour can be outlined in this process.

A first typology is represented by the United Kingdom, where the early liberalisation process stimulated and oriented European policy.

A second typology groups some countries, such as France, where the Commission's policies were shared and adapted to a content and timing that were more suited to the French context, in order to avoid negative effects on the national provider.

The third typology is represented by countries, such as Italy, where uncertainties, difficulties and delays in adopting the Commission directives give the impression of implementing them without sharing them, or of being incapable of taking advantage of the consequent opportunities.

A progressive identification of distinct interests, perspectives and roles resulted from the intense and dynamic interaction between the different players. The last to come on the stage, although still with quite a weak voice, are the regional and local authorities – both as users and as policy players – and the private users.

With regard to the regions, it can be seen that in those countries with a high degree of institutional and administrative decentralisation, regional authorities are more aware of the relevance of the sector, are capable of formulating a strategy, and have the will to intervene in the sector. An example is Spain, where some regional authorities developed their own strategies and objectives as the basis for negotiations with the national provider and for promoting the use of networks and services in those areas for which they are directly responsible – healthcare, education, training, public administration. Another example is Germany, where the Länder have for some years played an active institutional role in defining the rules and planning guidelines for the sector's policy, as well as promoting new infrastructures and services in their territory.

In Italy, the regional authorities have shown a level and capability of initiative that is lower than could be expected on the basis of their degree of autonomy. Various factors are responsible for this. Namely, the low dynamism of the sector, which delayed the development of an awareness of the potential of telecommunications; the difficult relationship between the state and the regions over the management of those areas where their responsibilities overlapped slightly – in Italy, the regional authorities do not have direct responsibility for telecommunications, but this sector is becoming more and more critical for those functions that are typically regional, such as economic and territorial planning; the well-known difficulties created by the prolonged instability of the political framework and the consequent vacuum in the public administration.

With regard to users: on the one hand, service providers are devoting increasing attention to their peculiar needs – although this is mainly focused on users who produce relevant traffic; and on the other hand, European Commission initiatives and policies are being addressed to users' needs and are trying to stimulate a more active role by them, involving their associations and individual users in pilot projects and in demonstration activities. Users' associations are playing only a small role both in negotiations with network operators and service providers, and in the dialogue with the regulatory authorities at the European and national level.

The rich and dynamic debate within the regions is supporting the launching of actions designed to stimulate, qualify and aggregate the effective and potential demand for telecommunications services by economic actors who are weak, either because of their small size or because of their geographical location.

Discussion of the geographic diffusion and coverage of information and communications technologies, in terms of the accessibility and affordability of networks and services, is still at an embryonic level. It deserves more attention both in terms of analysis and in the formulation of amending strategies, especially to provide guarantees of geographical diffusion of competitiveness and its benefits. In those areas and for those users who are not attractive enough for new operators, there is the real risk of the persistence of a monopoly, exacerbated by less protection for users.

Signs of change, besides those mentioned above, are also apparent in the behaviour of telecommunications operators. Two of the more recent signs are the Europeanisation of some national telecommunications operators and the inter-sectoral agreements signed by companies operating in different industries in the communications sector.

Within a wide process of internationalisation, there are signs of an increase in the size of the market for each of the national operators who now look at Europe as their operational area. Among the more relevant initiatives of this kind, it is worth mentioning the agreement between France Télécom and Olivetti (1995) designed to create a new telecommunications operator in the Italian market, and the agreement between Cable & Wireless and Veba (1995) for the German market. Besides these agreements, a range of other commercial initiatives is taking place in almost all the European Union countries.

At the same time, the first attempts at linking telecommunications operators and media companies are taking place. One of the best known, that which was planned by France Télécom, Deutsche Telekom and Bertelsmann for joint entry to the new market of multimedia products and services, was stopped by the European antitrust authority. Another that is worth mentioning, is the negotiations – at the time of writing – between British Telecom and Fininvest, to become the third mobile-phone operator in Italy, using the existing network of a broadcaster of Fininvest.

These changes, initiatives, attempts to set agreements and expand the operational field of activity are certainly the result of a number of factors. They seem to be accelerated by the need to reposition in the European context before the deadline for the complete liberalisation of services and networks, which fixed for 1 January 1998. Each of the chapters in this volume contributes to the definition of this complex panorama and analyses in depth the themes that represent the debate in the telecommunications sector in Europe.

Convergence

I The Challenges to Regulation in Europe

Telecommunication:
New Dynamics and Driving Forces
A. Bianchi and G. Richeri (Eds.)
IOS Press, 1996

Chapter 1

Convergence and trans-European networks: some policy problems

Jean-Claude Burgelman and Pascal Verhoest
Studies on Media, Information and Telecommunication,
Free University of Brussels, Belgium

Abstract: This chapter offers a critical examination of policy and especially its assumptions on the superhighways and the convergence of broadcasting and telecommunications. Our diagnosis of current European Union policy is that it is too oriented or led by three criteria – competition, short-term return on investment and technology push – which make the realisation of the EU infrastructure-policy goals highly questionable. This is confronted with what we think is a better way to deal with convergence and the ambition to create superhighways in Europe ('new deal' type of regulatory policy, long-term objectives and user-led annex-content oriented).

1. Introduction

European policy-making with regard to trans-European networks (TEN) is based, among other things, on two assumptions. Firstly, that the growing demand for new services, home-entertainment, multi-media applications and teleworking requires pan-European convergence between telecommunications and broadcasting. Secondly, that the building of the proposed trans-European network can only be achieved in a competitive environment [1].

Convergence is a complex matter. It is, at present, difficult even to give a precise definition of what is understood by it, let alone answer the question as to where convergence will lead and what the related social problems will be. However, a synthesis of the mainstream literature on this matter would read as follows.

Technological development allows integratioh of networks. Digitalisation leads to the blurring of the physical/material distinctions between them, since every carrier (fibre, copper, coax, spectrum) is a potential candidate for transmission. 'The mother of all changes is digital', the American press stated a while ago, thereby suggesting that digitalisation would imply abundance instead of scarcity.

There is, however, little clarity on the nature of the new communications (dis?)order that will come out of the current evolution. At best, one reads that the hope is that the new communications order will lead to a better society and more economic prosperity. Many of the accounts on the subject are only supported by notions such as 'we have good reason to believe that the consumer wants this or that' or 'there are strong feelings about the communicational needs of the people', etc.

The behaviour of the main actors involved (carriers, software industries, media and users) is difficult to categorise and quite unpredictable. The only emerging pattern seems to be that many act so as 'not to miss the boat', although they may not be clear what they may gain. Coherent policy patterns are therefore difficult to detect, not least because many of the actors in the field carry out strategic 'window dressing'.

As this paper puts forward, regulatory problems are important and may require new approaches. These problems range from privacy and data protection, respect for general laws on morality and rules regarding fair trading, to problems dealing with cross-ownership and concentration. From this perspective, it is tempting to put forward chaos theory as a final 'recommendation' and as a possible 'explanation' for what is going on, leaving it to the market to find a way out in the hope that some order results.

However, instead of regarding convergence as a black hole, one can try to understand the phenomenon. Past experiences with regard to communications infrastructure development on the one hand, and the dynamics of communications media/technologies on the other, puts the problem in a clearer perspective and may help to formulate the principles necessary for a good policy in this field. This paper discusses some economic and political policy aspects of the development of information and communications services. It does so by focusing on European Union (EU) policy, or intended policy, with regard to convergence and trans-European networks.

One specific problem needs to be mentioned in this respect. The amount of literature (academic, policy and consultancy) is relatively abundant but difficult to judge in terms of factual accuracy and empirical reliability of the analysis presented. It should be noted that hard, cross-national and cross-sectional data, giving a reliable overview of all actors concerned, are hard to find. Even where figures are available, interpretations can be misleading.

For example, when discussing the supposedly bright future of the cable industry and the importance of this evolution, many accounts refer to this market in terms of homes passed (which are many) but only very seldom in terms of actual connections (which are much fewer). Another example deals with the real (instead of potential) European audience reach of the pan-European satellite. Reliable figures are hard to find. For Belgium, the audience of these channels is, cumulatively speaking, considered to be less than five per cent. These specialised pan-European channels are one of the few successful applications of narrow-casting up to the present. They are considered to be an important indicator of the success of communications highways, since the assumption is that the demand for narrow-casting will rise dramatically. It is therefore important to have the right figures at hand.

The same applies to the multi-media market: reliable figures are, to our knowledge, not available. The most recent report speaks of 311 available CDi titles (Bouwman, p 93)[2] with 200 more in production – a significant amount as such, but it is questionable whether this is sufficient to build a communications highway. In sum, it appears that on most of the subjects (TV entertainment, telework, multi-media) on which the EU bases its hopes of building information highways in Europe, there is a lack of transparency of the market structure and a lack of coherent data. At this point, at least one urgent need for action in this field can be put forward: the need for reliable comparative data for the communications sector at large (production, distribution, consumption).

Due to the 'economics of research' (lack of money and time) it is impossible to deal with this scarcity in this paper. It is also impossible in a few pages to present a totally balanced, well-documented and academically-sound report on such a complex problem – if such is possible at all. This paper, therefore, neither goes into great detail nor presents all the elements of the debate.

2. Some lessons from recent history

The American discourse and policy goals with regard to the National Information Infrastructure (NII), its potential and its future can, to a large extent, be compared with the discourse held by the EU at the beginning of the 1980s, with the launch of programmes like RACE and ESPRIT [3]. The difference is that fifteen years ago it was forecast that all the wonderful new techniques would enter the home in the mid '90s, whereas the present forecast speaks of 2010.

The goals put forward by the NII are strikingly similar to what the RACE programme said it would deliver, and indeed tried to deliver for many years, ie, creating a broadband network because of the assumed (urgent) need for all kinds of new communications services. It is fairly commonly agreed that, for many reasons, RACE did not fulfil expectations, not least because RACE was built on the assumption that one type of technology would make it, whereas it was precisely the evolution of technology itself which made the original broadband idea less attractive.

The history of communications technology shows that the emergence of a new technology often goes hand in hand with a lot of hype ('this will revolutionise the way we live'), mystification ('this was what mankind was born for...') and ideologically-biased statements ('the wired society, where all will have equal opportunities, will be created'). When radio surfaced in the industrial world in the 1920s, the predictions about the impact it would have on our society were almost the same as those currently being attributed to NII. Most of the time, as explained elsewhere [4], forecasts and expectations of any new communications technology are inspired by the desired evolution not the reality.

In implementing the present policy on the subject of the trans-European networks ([1], the Fourth R&D Framework and so on) the EU could benefit from looking at the past, for instance, the broadband 'failure' of RACE in terms of its original purpose. The minimal lesson to be learnt here is that technology-specific policies are often problematic.

3. Elements of convergence

Convergence is technically enhanced by digitalisation or compression (of content) or both. It is expected that in the near future almost everything can be transported, from point A to whatever point B, in any physical way possible (copper, coax, fibre, spectrum). All existing communications infrastructures will be able to integrate voice, data and pictures interactively. Today, however, convergence still deals with two relatively distinct communications spheres.

Table 1.

	Telecommunications	**Broadcast**
pattern	point to point/individual	point to multi-point
time lag	free	fixed by sender
use	individual	collective
financing	per unit	subscription
regulation	freedom of communication	content regulated
perceived as	economics	culture
industry	telecommunications organisation	broadcasting organisation

3.1. The services

Basically convergence enables any type of existing service (for example, telephone, TV), enhanced telecommunications (for example, caller ID) or new TV service (for example, video-on-demand), or a mix of both (video-conferencing, home-shopping). However, given that many of these services are already being supplied, what is likely to happen at first is that telecommunications organisations will start to offer broadcasting services, and vice versa (as is the case in the UK).

More specifically, the broadcasting market is regarded as the one with the most immediate growth potential for developing new services (video-on-demand, narrow-casting, closed-user TV, etc) Many analysts suggest that the outcome of competition between the different networks (cable, telephone and so on) would begin with video-on-demand. Convergence makes it possible for a one-way mass medium (one sender to many receivers and a non-manipulative time schedule) to become distributed via a two-way medium. Or, to put it another way, convergence makes it possible for all media to be consultative (to be consumed at any time) or even to be conversational (interactive).

The next stage then would be multi-media applications. In its real form (integrating text, data, voice, pictures in an interactive way) this is considered to be the ultimate in convergence, as it integrates production and distribution of hard and software in one service. At this point it might be useful to remember that many accounts use the concepts of multi-media and (network) convergence in the same way. However, multi-media-type applications, like CDi, are only one way of offering an integrated service (and in fact do not necessarily need to be supplied via a network).

3.2. Technical potential and network evolution

The technological convergence dream involves fibre-to-the-home, because of its so-called unlimited capacity. This was the initial aim of RACE: one broadband ISDN network in Europe, offering all communications services. However, apart from some selected trials, not much has happened to date.

Alternative scenarios for the future implementation of a broadband network do exist. The first, ISDN/FTTH (integrated services digital networks/fibre-to-the-home) draws on the original ideas. The second consists of upgrading existing networks (switches and cable) and using new carriers (for example, spectrum). Many analysts believe, given the level of investment (in cable and switching) sunk in broadband ISDN (Nippon TT abandoned its $360 billion plan to get FTTH by 2015), that the second scenario will take over in coming years.

It is important to note that, in either or both scenarios, the 'capacity' of a network becomes less and less relevant (the highest costs are in the building of the network). However, the cost structure of such networks remains extremely difficult, not to say impossible, to work out [5] [6] [7] [8]. This is an essential problem and will have specific relevance for users. If the cost of renting a movie is 5 ECU (video-on-demand) this will involve use of the network for two hours. Is the implication, therefore, that one can also phone for two hours, for 5 ECU, on the same network? The answer could lie in dynamic pricing, according to demand. But would not that be the same as what the old monopolistic telecommunications operators (TOs) did in the past? Value-oriented pricing can lead to determining prices in an arbitrary way, as present day critiques of the TOs' pricing policy imply, hence the European Commission's (EC) recommendations for orienting prices towards costs.

With regard to media, production costs for mass media are expected to drop dramatically due to the digitalisation already mentioned (no longer requiring mass audiences, for example). 'A bit is a bit', it is said. But when everything is a 'bit', seemingly unresolvable questions with regard to copyright, intellectual rights and so on, will emerge.

Finally, as the possibilities for trans-border data flow would be unlimited (for instance, the accounts of Swiss Air being processed overnight in Bombay), geographic boundaries become irrelevant. It is, technically speaking, perfectly possible to dial up a video-on-demand service in the US and view it in Europe. But how will national or regional policy-makers react, when they lose control of their economic fabric?

3.3. *Actors/stockholders*

Assuming the EU will open up the possibility for full competition on infrastructure, the economic actors or potentially interested parties in convergence are multiple:

- all possible carriers (TOs, cable companies, satellite operators, and potential new entrants from the public utilities sphere, like the gas and water distributors, the railways, electricity providers, etc);
- software companies (to provide intelligence in the network, on the screen and in-between: cruising or navigation services; see the recent initiatives by Oracle, Microsoft (*Communications Week* 17.1.94));
- media 'content' industries (broadcasting, film, publishers etc);
- telecommunications-equipment manufacturers;
- consumer electronics (linked to TV or to video games);
- computer manufacturers (PC industry).

Although it is difficult to determine the importance those actors (will) have in the field, it seems that both telecommunications organisations and the broadcasting industry, as content provider, are at the core of the convergence business.

Telecommunications organisations are one of the main drivers behind convergence, not least because the switched (telephone) network is still being only poorly used. The average usage of a telephone line is not more than 30 minutes a day. For residential users in the UK it is only between five and seven minutes a day (*Financial Times* survey October 1993). This leaves the network largely under used, especially outside peak hours.

On the other hand, the interest for the broadcasting market is easily explained by the fact that watching TV is one of the main human recreational activities. In other words, and in terms of user demand, no new user habit needs to be established. This 'certainty of demand' is very important, given the financial effort required and the competitive environment within which the networks will have to operate (see also next section).

The financial stakes are indeed extremely high. For example, to offer movies, home-shopping, games and video-on-demand via the switched network in five cities (to 500,000 homes) US West says it will need $750 million over the next two years (AP 4.1.94). Notwithstanding the huge amounts of capital, the EU, like the Clinton administration, wants the networks to be financed largely by private capital [1]. Competition policy of the EU, therefore, wants to keep barriers to market entry as low and as transparent as possible, hence open network provision (ONP) and the like.

This would mean that actors wanting to invest in future networks and services, would only be able to do this in competition with other potential actors. High investments costs require economies of scale or scope or both in order to recoup the capital invested, hence the need to create trans-European companies. Even then, success cannot be guaranteed even though demand may be high. Both the Danish and French TOs in the 1980s built highly digitialised nation-wide cable networks but are still losing money. Both are looking for new services which will appeal to the users.

4. Possible consequences of convergence

Asymmetric regulation is needed to improve competition in this field, to favour new entrants and disfavour dominant actors (see Oftel's policy towards British Telecommunications with regard to cable). In the UK this has resulted in US cable and telecommunications operators entering the UK telecommunications business 'via the cable'. In March 1993 they accounted for 143,000 subscribers; only ten per cent of which were business users. However, to come to the pan-European networks it is crucial that the key actors, like the telecommunications operator, are able to supply new video services as well. This will act to reinforce their dominant position.

4.1. Consequences of convergence related to the actors involved

As stated before, cost structures of telecommunications networks are not transparent, and moreover, dominant TOs may 'win' by maintaining the status quo [9]. TOs may not be willing to take the risk of being degraded to a simple 'carrier'. They could rather want to move into horizontal as well as vertical differentiation: telecommunications, cable, software, consumer electronics and content producers would thus become interlinked. This would mean the emergence of a few oligopolistic communications conglomerates [10].

To be competitive in an environment of convergence and multi-media application, a corporate actor must seek to control the four domains of the communications process (production, distribution of hardware and software, including content). This will stimulate the formation of pan-European and multi-media groups and will raise specific questions with regard to legitimating a specific policy on pluralism and diversity.

A scenario of full competition will, given the amount of investment at stake, most probably be very difficult to sustain. Investment in pan-European highways must be long term and requires collaboration. Therefore, the emergence of conglomerates is likely, probably with telecommunications companies at the head because they are the oldest companies involved and are much less indebted or are quite simply the richest.

Cross subsidisation – an essential element in the EU's competition policy – will have to be extremely carefully looked at also. Telecommunications companies can very easily drop their prices when competing in television, for example, with cable companies, by using their telephone revenue. Or vice versa, for example, when the local UK cable companies decided in January 1994 to join forces (the beginning of a duopoly?), drop their prices for telephony and cross-subsidise from their TV revenues (*TechEurope*, January 1994).

Moreover, if new entrants, like the electricity companies or water companies, come in as carriers – which they will be entitled to do if not already doing so (see Général des Eaux in France) – new problems of cross-subsidisation (from electricity revenues to communications) may arise. How will this be solved given the pricing problems already mentioned?

As stated earlier, the video-on-demand market is considered to be crucial. This is because it is assumed that there is potential demand for such services and because it is, in the short-term, the most profitable-looking application. It is believed that whoever controls this market will win the competition. Although no one is really certain of this market [11], many actors are forming strategic alliances with the film and broadcasting sectors. In other words, and looking at the US experience, one can expect collaborative agreements with regard to mergers in this field (telecommunications-cable-broadcasting) in the very short term.

4.2. Consequences of convergence related to regulatory issues.

Although radio was originally a byproduct of telegraphy, broadcasting and telecommunications were very quickly regulated as separate entities because they developed as distinct services. Convergence brings these two worlds together in one or more (depending on interconnection, etc) physical infrastructures.

It becomes possible to combine broadcast and narrowcast services in with a return channel, thus enabling all sorts of segmented broadcasting (closed-user, pay-per-view, video-on-demand, etc). In regulatory terms, this means that the old discourse on broadcasting – that it was a scarcity and hence subject to control in the public interest – is no longer valid.

The question arises as to how to reconcile broadcasting and telecommunications regulation, which is based on the concept of general public reception and the protection of cultural and political values on the one hand, and on the other telecommunications regulation, which is based on individual exchange – the regulation of message carriage instead of message content [12].

This also raises questions, such as what is the nature of telecommunications-delivered content? Is a 'porno' database, delivered by a TO, subject to the normal rules of media control or to those of the TO, where there is much less content control? If a racist party starts a closed-user broadcast channel, can it be held accountable for it, as is currently the case?

The privacy problem – as clearly shown by the consumer resistance in Germany to the FTTH plans of Deutsche Telekom [13] – should not be underestimated either: the digitalised registration of watching habits, consumer behaviour via tele-shopping and home-banking opens up extremely interesting marketing perspectives that may be not desired by everyone.

These examples clearly illustrate that regulatory principles based on the mechanisms and patterns of transport are no longer adequate and need to be re-conceptualised. New techniques will also blur the distinction between closed-user TV and subscription TV. What is the implication here? What if, for example, the service potentially reaches more than one, two or twenty per cent of the audience, when it is offered free? If one wants to regulate at all, a new definition of mass media will be needed.

A re-definition of the principles that should guide telecommunications is also necessary. It must be borne in mind that universal service was the only guiding principle for telecommunications and was the result of a compromise between telecommunications operators (be they state or private) and political authorities: in return for a monopoly (be it state or private) the latter imposed universal-service obligations. What does universal mean in an environment where pay-per-service will be the driving principle?

4.3. Consequences of convergence related to demand

As stated previously, the question of the demand, for Europe as well as for the US and whatever broadband scenario followed [14], is the crucial point of the whole debate. Given that the European networks will have to be established with private capital, the level of demand for new services will, to a very large degree, determine the available capital for investing in the networks, upgrading, etc. This is reinforced by the fact that no potential carrier is, in a competitive environment, certain of what his revenues will be, as the old monopolistic – whether state or otherwise – carriers used to be.

Therefore, the question is, whether there is a significant demand for new services and, if so, will it be large enough to support the necessary investment? Will consumers, before the end of this century, have access to 300 TV channels? Will they be using video-on-demand? Will they consult the libraries of the world from the screen and, in between times, do some telework and tele-shopping? – as forecast by the French Minister of Culture, A. Carignon, recently (Idate 1993).

As pointed out in the introduction, it is very difficult, due to the lack of accurate data, to give clear answers to these challenging questions. It is, by the way, remarkable that on a such a crucial point, so little has been invested in research into what users really want, as opposed to what the marketing manager wants them to want. Much of the literature lays stress on the fact that no one knows what the real demand will be and that few methods exist to determine it [15]. Nevertheless, looking at past successes and failures in communications technologies, some general remarks can be made here.

First, there is no reason to believe that in the immediate future an explosion in demand will take place. Almost all initiatives with interactive TV failed (in the US as well as in Europe): the videophone was a failure, segmented pan-European TV – after more than 15 years of trial and error – has attracted only a very small fraction of the market (less than five per cent); there is, at world level, at maximum, room for only two or three CNNs and the VCR.

Second, there is a clear distinction between the adoption process of information media or new telecommunications services *strictu sensu* and more traditional media (such as broadcasting, whatever the delivery system may be). There is no reason to believe that while some individuals cruise on Internet, they will do the same in broadcasting.

The notion of 'acquisition substitution' is essential in understanding the demand and adoption process: will a new service offer the same function at a cheaper or more efficient rate? Cable offered more TV of better quality. The fax offers better written interpersonal communications at a higher speed (and thus more cheaply) This acquisition substitution is different when the new service has to be adopted in the professional or residential sphere. In the professional sphere, every new technology that improves efficiency, profit-making, and so on, is likely to be adopted (given the substitution factor). However, it should be remembered that only rational elements come into play here (going to a conference is more fun than video-conferencing). In the residential sphere, the acquisition substitution is also determined by social and entertainment values. In other words, what will video-on-demand substitute for? Will the consumer be willing to pay for it, given that he or she has a VCR, when renting movies is quite cheap, and so on? And finally, what does it add to the multitude of channels already available? According to some sources, the trials with video-on-demand in the US revealed a demand of no more than two or three movies a year.

Third, introducing radical new communications technologies, ie, those which do not directly substitute an existing method of communications, needs much more time, social experimentation, careful organisational planning, and so on. Here, the Minitel is the best example. France still has by far the largest usage rate of videotext [16]. The French strategy was also the most specific: free Minitels, free access, long-term investment by France Télécom, cross subsidisation, the Minitel rose, and so on.

The VCR is another good example in the residential sphere. It took several years to take off and it gained ground against the video disc which was launched at the same time, because of the fact that the VCR offered something the video disc could not – viewing autonomy towards the channels and thus possibilities for time management (although it was first seen as an alternative for Super 8).

Taken together, these few principles can be summarised as follows:

- creating demand in the short run, makes most sense when it is applied to services that have an immediate substitution value to the business or residential consumer;
- only in the long run and if accompanied with a lot of 'social engineering', does the development of radical new services make sense;
- this process differs substantially when applied to the residential or to the business user;
- there is very little information on user behaviour.

4.4. *Some remarks on the assumption of a highly segmented broadcasting market*

The idea of 300 channels being available with highly specialised and specifically targeted productions, combined with video-on-demand and so on, might be challenging, but raises some questions. Some critical reflections can be formulated on the expected explosion of channels (see also the remarks just made with regard to substitution).

Research shows that over recent years the total amount of TV consumption stabilised in Europe (three hours a day). This means that new services will have to substitute existing viewing habits [17]. Will consumers pay for something (video-on-demand) which they already have? The VCR was no challenge to the broadcasters (who compete in a segmented market by showing even more of what people want) but to the film theatres. How user-friendly is video-on-demand likely to be? For instance, will the viewer be able easily to browse through a (screen-based) catalogue?

Do consumers really want all these screen-based applications integrated into their TV? If so, why was it that the breakthrough of computer games (Sega and Nintendo have a penetration of 30% in the US residential market (Van Trier, p 9)[18]) occurred only after they were developed as stand-alone applications.

Who is going to finance a segmented offer in Europe? It took CNN more than ten years to reach break-even and it is highly questionable whether the market can support more than three of this type of news channel world-wide. How is advertising going to respond to a segmented market? Is there enough advertising revenue to support it (given the fact that the largest TV advertisers are the Procter and Gamble-type, producing consumer goods)?

The economics of TV are not comparable to those of the press (although the analogy is often made): it requires the same investment to make a TV programme for one viewer as for one million. Is the European market large enough to allow specialised TV on a subscriber basis (500,000 subscribers are required in the US according to B. Diller, the president of the home-shopping channel QVC)? Or, more precisely, is it homogenous enough? There are, for instance, enough hunters in Europe wanting to see a hunting channel, but will the hunters of Spain be content to see what is of interest to them produced in Denmark and broadcast in English?

Cross-border TV viewing, essential to allow a pan-European segmented offer, is very low in Europe. In Flanders, for instance, (6,000,000 inhabitants), where the audience has had more than 20 channels at its disposal for the last 15 years, and knowledge of French and English is wide, more than 70 per cent watch the Flemish channels, ten per cent look at the Dutch channels (same language), ten per cent watch VCR and ten per cent watch the rest (French, BBC, etc) After several years of promoting itself, Filmnet, the pay-TV channel, has in this environment of abundant choice, managed to get only 135,000 subscribers and this includes a sharp increase in recent months due to the programming of soft-porn videos.

How will the viewer make a choice between 200 channels? The solution put forward is that someone will offer to make the choice for them (navigator systems). For example, 'I want to see so much of so and so and navigator will do this for me and bring me half an hour of soap, fifteen minutes of news, a movie, and so on'. Although possible, isn't this re-inventing the generalist broadcaster?

5. The specific character of the EU

The problems that can be perceived for the building up of trans-European networks are very similar to those which caused the failure of RACE [19]. They concern the conflict between the will of the EU to install fair competition, open the market, abolish monopolies, and so on, (the ambitions of DG IV) and the need for at least a stable environment with guaranteed revenue to build the TEN (the ambitions of DG XIII). On top of that, the concept of TEN relies heavily on the broadcasting and media sector. This means that one more different regulatory world (that of the media, but also of DG X) comes into play. In sum, three areas of EU regulation need to be linked with each other: competition policy, telecommunications policy, media policy. Is a unified European regulatory approach the answer? [20]

The Green Paper on pluralism and the media does not address many of these issues, since it only looks at cross-ownership between broadcasting and the press. Issues not taken into account are vertical integration and cross-ownership with industries other than broadcasting or press (TOs, software companies, etc).

The concept of TEN is an infrastructure concept, whereas the services to be provided on that network are software (programmes). If broadcasting and related services are so important for the take off of the 'converged' TEN, Europe will certainly have to develop its strength, internationally speaking, in the software/content area. It is paramount that European policy-makers can develop this strength, without having to consult the member states over every single policy measure, as is the case now. The so-called cultural matters are, however, a core concern of the member states, as is illustrated by the way they were left out of the GATT. This illustrates the need for a 'pact on TEN' between the EU and the member states.

Table 2.

More 'realistic' policy	Current EU policy
'guaranteed' revenue	agitated market
'new deal' type of policy	competition
user-led	technology push
content-oriented	the medium is the message

A specific and global EU policy with regard to convergence based on what we know of the history of innovation in communications and networks, and assuming that the concept of TEN is a realistic one, could include the following broad policy principles, which are needed to make the TEN a success. First of all, there is a need for a stable economic environment to develop the networks and create demand. Secondly, this necessitates an active and 'superimposed' public policy. Third, a user-driven (content-oriented) and user-specific (users are not all the same) policy is to be recommended.

If a study is made of two of the most successful initiatives in the last fifteen years with regard to the information society (Internet and Minitel) it will be found that they are the product of a strong central commitment, a coordinated, long-term and dedicated financial effort (enabling cross-subsidisation), and a stable regulatory environment to encourage a chosen direction of technological evolution, while at the same time allowing opportunities for users to experiment (free access) and kick off the service.

6. Conclusion

It remains an open question as to whether convergence and the move towards an integrated network will occur at all the levels discussed (actors, networks, services). Recent reviews [21] [22] in a number of European countries revealed that one cannot speak of a breakthrough on all the levels discussed here. At the level of the networks, the vital questions of who will finance FTTH or whether upgrading is not a better alternative or both, remain unanswered. At the level of the services, the key question still is, to quote a recent conference of captains of the TV/Telecom industry in the US on this subject: 'Will Mrs Jones (the American Mrs Dupont) push the TV button and explore the multi-media world behind it or not?' (*Le Monde* 2.2.94). These unresolved problems confirm the conclusions of the Tecknibank study [23] that a strong stimulation for both infrastructure implementation and usage innovation will be needed if the EU wants the TEN to be a success.

However, what is undoubtedly the case is that convergence is taking place at the level of the actors, thereby creating integrated communications industries (hard and soft, production and distribution) and generating all kind of problems with regard to cross-ownership, pluralism, diversity, support mechanisms, media policy, and so on. On the other hand, the digitalisation of communications is also advancing, generating problems with regard to ownership, intellectual property rights, copyright, control, privacy, and so on. In sum, convergence necessitates a new economic, regulatory and legal approach.

Notes and references

[1] Commission of the European Communities, Growth, Competitiveness, Unemployment: White Paper. Commission of the European Communities, Brussels, 1993.

[2] H. Bouwman, De vraag naar consumententoepassingen, *Informatie & Informatiebeleid*, 4, 1993, pp 92–96.

[3] M. Joosten, De verloren integratie. In P. Slaa & J. C. Burgelman (Eds), Verglaasde steden. O. Cramwinckel, Amsterdam, 1994.

[4] J. C. Burgelman, Verhoopte, voorspelde, reële en te verwachten tendenzen in de ontwikkeling van communicatietechnologieëen. In H. Bouwman en N. Jankowski (Eds), Interactieve media op komst. O. Cramwinckel, Amsterdam, 1989.

[5] N. Garnham, Telecommunication Policy in the UK. In N. Garnham (Ed.), Capitalism and Communication. Sage, London, 1990.

[6] Mediaraad, Informatie: tot (w)elke prijs? Mediaraad, Den Haag, 1993.

[7] G. J. Mulgan, Communication and control. Networks and the new economies of communication. Polity Press, London, 1991.

[8] M. Joosten, Restructuring telecommunications tariffs in Europe. PhD manuscript, 1993.

[9] P. Vehoest, A Decade of Liberalisation in UK Telecommunications: a policy assessment. Science Policy Research Unit, University of Sussex, Brighton, 1994, p 50.

[10] R. Mansell, The new telecommunications. A political economy of network evolution. Sage, London, 1993.

[11] L. L. Johnson & D. P. Reed, Telephone company entry into cable television: an evaluation, *Telecommunications Policy*, 2, 1992, pp 122–134.

[12] P. Verhoest & J. C. Burgelman, Les services publics de la communication en Belgique, *Réseaux*, 3, 1994, pp 67–97.

[13] D. Garbe, Social compatibility of telecommunication technologies, *Telecommunications Policy*, 11, 1992, pp 646–656.

[14] M. J. C. Elton (Ed.), Integrated broadband networks: the public policy issues. North-Holland, Amsterdam, 1992.

[15] J. Leyten, T. Weijers & H. Schaffers, Informatie en communicatiediensten en de communicatie-infrastructuur: Opties voor beleid. TNO/STB, Apeldoorn, 1991.

[16] Impact, Main events and developments on the electronic information services market. Annual Report 1992. Commission of the European Communities, DG XIII/E, Luxembourg, 1993.

[17] R. Silverstone & L. Hadden, The individual and social dynamics of information and communication technologies: present and future. Paper for the RACE project 2086. CICT/Science Policy Research Unit, University of Sussex, Brighton, 1993.

[18] G. Van Trier, De markt voor wetenschappelijke en technische informatie in de Verenigde Staten, *Informatie & Informatiebeleid*, 4, 1993, pp 8–16.

[19] G. Fuchs, ISDN – the telecommunications highway for Europe after 1992, *Telecommunications Policy*, 12, 1992, pp 635–645.

[20] J. P. Simon, Towards a unified European regulatory system? International Telecommunications Society Paper, 1993.

[21] E. Bustamente, N. Garnham & J. M. Salaun (Eds), Téléphone et télévision, enquête sur une convergence européenne. CNET/Réséaux, Paris, 1993.

[22] P. Slaa & J. C. Burgelman (Eds), Verglaasde steden. O. Cramwinckel, Amsterdam, 1994.

[23] Technibank, The impact of the accelerated adoption of broadband communications on the European Community economy. Commission of the European Communities, DG XIII, Brussels, 1993.

Analysis, Performance of the telecommunications sector up to 2010 under different regulatory and market conditions. Final report to the European Community, 1992.

J. C. Arnbak, J. J. Van Cuilenburg en E. J. Dommering, Verbinding en ontvlechting in de communicatie. Een studie naar toekomstig overheidsbeleid voor de openbare electronische informatievoorziening. O. Cramwinckel, Amsterdam, 1990.

J. C. Arnbak, J. J. Van Cuilenburg en E. J. Dommering, De openbare elektronische informatievoorziening. Achtergrondstudies voor het project lange termijn Mediaplanning. O. Cramwinckel, Amsterdam, 1991.

D. J. Atkin, Adoption of Cable Amidst a Multi-media Environment, *Telematics and Informatics*, 1, 1993, pp 51–58.

F. Bar & M. Borrus, The future of networking in the US. Berkeley Roundtable on the International Economy. Final draft, 1992.

J. C. Burgelman, Van massa communicatie naar nieuwe communicatiepatronen. In J. Bardoel en J. Bierhoff (Eds), Media theorie. Wolters Hoordhoff, Amsterdam, 1992.

J. C. Burgelman & C. Pauwels, Growing convergence between broadcasting and telecommunication: policy problems at the level of the Commission of the European Communities, *Telematics & Informatics*, 3, 1991, pp 135–141.
J. C. Burgelman & C. Pauwels, Audio-visual and cultural policies in the small European countries: the challenge of an unified European television market, *Media, Culture and Society*, 1992, pp 169–183.
A. Calabrese & D. Jung, Broadband telecommunications in rural America: emerging infrastructures for residential services, *Telecommunications Policy*, 3, 1992, pp 225–248.
J. P. Chamoux (Ed.), Deregulating Regulators? Communications for the 90's. European Communication Policy Research Series. IOS Press, Amsterdam, 1991.
D. A. Dunn, Cable-Telco Competition, *Telematics and Informatics*, 2, 1992, pp 73–86.
European Publishers Council, The emergence of a multi-media industry in Europe. Press conference, Brussels, 1993.
N. Garnham & G. Mulgan, Broadband and the barriers to convergence in the European Community, *Telecommunications Policy*, 3, 1991, pp 182–208.
Y. Gassot, La câblo-téléphonie au Royaume-Uni, *Communications & Strategies*, 9, 1993, pp 119–125.
R. A. Gershon, Telephone-cable cross-ownership: a study in policy alternatives, *Telecommunications Policy*, 2, 1992, pp 110–121.
M. Goodman, K. Lu et al, Telephone company entry into cable television: a re-evaluation, *Telecommunications Policy*, 2, 1993, pp 158–162.
T. W. Hazlett, Rent-seeking in the Telco-cable cross-ownership controversy, *Telecommunications Policy*, 5, 1990, pp 425–433.
W. Hulsink & G. J. Kemme, Van publiek monopolie naar zelfregulerend kartel? Institutionele aspecten van de openbare electronische informatieverzorging. Rotterdam School of Management, Report 106, 1992.
Informatie & Informatiebeleid, themanummer 'Multi-media tussen hope en hype', 4, 1993, pp 14–97.
J. Liebowitz, Promising Applications of Multi-media, *Telematics and Informatics*, 2, 1993, pp 149–156.
D. Lindsay, When cultures collide: regulating the convergence of telecommunications and broadcasting. Policy Paper 29. CIRCIT, Melbourne, 1993.
G. J. Mulgan, The myth of cost based pricing, *Intermedia*, 1, 1990, pp 21–27.
OECD, Convergence between communications technologies: a policy review. OECD, Paris, 1991. Restricted document.
OECD, Seminar on the policy implications of the convergence between communication technologies. OECD, Paris, 1991. Restricted papers.
PACE, Perspectives on advanced communications. Commission of the European Communities, DG XIII, Brussels, 1990, 1992.
V. Patel, Broadband convergence:a view of the regulatory barriers, *Telecommunications Policy*, 2, 1992, pp 98–104.
Y. Punie, A. Veller, P. Verhoest & J. C. Burgelman, The diffusion of telematic innovations: towards a user-led diffusion strategy. International Association of Mass Communications Research paper, Dublin, 1993.
J. L. Renaud, Conséquences de la convergence entre technologies de communications: le cas de l'Europe. OECD, Paris, 1990. Restricted paper.
J. P. Simon, L'Esprit des regles: réseaux et reglementation aux états unies. cable, électricité, telecommunications. L'Harmattan, Paris, 1991.
J. Van Beukel, PTT in kabeltelevisie: concurrende netten, *Informatie & Informatiebeleid*, 4, 1991, pp 19–24.
Warburg, Multi-media – myth and reality. Warburg, London, 1993.

Telecommunication:
New Dynamics and Driving Forces
A. Bianchi and G. Richeri (Eds.)
IOS Press, 1996

Chapter 2

Information highways: the European flavour

Jacques Arlandis
IDATE, Montpellier, France

Abstract: European initiatives are now explicitly bound to the deployment of trans-European networks, hopefully achieving the objectives of the European Union. In the telecommunications domain, this may be regarded as an inflection in the Union rationale. This chapter raises the question of the real basis against which a European information infrastructure policy could actually be implemented. It argues that such a policy is mainly bound to the steady deregulation process which has been set up in the last decade, so giving rather limited scope for the strengthening of any European industrial policy. It stresses the inertia of telecommunications policies, presents the European commitment to the information highways and discusses the possible adjustments to European policies.

1. Introduction

European initiatives are now explicitly bound to the deployment of trans-European networks (TEN), hopefully achieving the objectives of the European Union. In the telecommunications domain, this may be regarded as an inflection in the Union rationale, taking into account:

- the regulatory context which modifies or even limits the scope for intervention by public authorities, including the European Commission, in the field of telecommunications networks (growing liberalisation of services and infrastructures);
- the contestable success of the main R&D programmes launched by the Commission (ESPRIT, RACE ...) in the last decade, in particular with regard to the deployment of advanced communications networks and the stimulation of a truly European demand, in scale and scope;
- the necessity of providing an appropriate answer to the national information infrastructure (NII) initiative launched by the Clinton-Gore administration.

Faced with what could appear to be a dangerous metaphor ('the superhighway'), the policy-makers in Europe have tended until recently to take refuge in their usual scepticism. There was a lot of comment when the NII project was presented as a type of global operation under the control of a single body, utilising a perfectly homogeneous technology. Above all, it paid no heed to existing telecommunications services that would suddenly be made obsolete.

Whether European or not, the superhighway promoters find themselves up against the following questions. In what way are telecommunications infrastructures and services – which are nonetheless attracting growing investments by operators – unable to meet the demands of certain users? How can a guiding role be restored to the public authorities when deregulation and competition has meant dealing with multiple carriers and the rules through which they interact?

This chapter raises the question of the real basis against which a European information infrastructure (EII) policy could actually be implemented. It argues that such a policy is mainly bound to the steady deregulation process which has been set up in the last decade, so giving rather limited scope for the strengthening of any European industrial policy. It stresses the inertia of telecommunications policies (section 2), presents the European commitment to the information highway (section 3), and discusses the possible adjustments to European policies (section 4).

2. From the Green Paper to Section XII of the Treaty on European Union: the inertia of telecommunications policies

Linking the construction and development of European telecommunications services through the setting up of trans-European networks is certainly not a recent idea. The concept may be traced back to the early 80s. Being more or less successful, it made its way through policies endorsed by the Community for about ten years. Such a long period, one formal result of which will be the drafting of Section XII of the Treaty on European Union, is the reason why we talk of inertia. The possibility of its realisation arose from the 1993 White Paper in which the Commission responded to the NII announced by the US administration which was elected to the White House at the end of 1992.

The Commission's Green Paper proposed partial deregulation and a weak industrial policy. A rapid diagnosis is required to put a policy centred on the problem of trans-European networks into perspective.

2.1. Liberalisation-harmonisation: the foundations of a policy of controlled deregulation

Interpretations of European telecommunications policies are of course available. In particular, they are a chronicle of proclaimed deregulation. Examination of the recent debate on the review of the European policy provides some understanding of the issue, although by 1992–1993 no real assessment had been made of the policy followed since 1985.

The announcement of the disappearance of monopolies by 1998 gave concrete expression to deregulation policy. In fact the review provides an effective and objective evaluation of the two directives in question: the liberalisation of services and the open network provision (ONP) which were introduced together in 1989.

The Green Paper of 1987 outlined a doctrine based on a European consensus, that of the smallest common multiple. The harmonisation-liberalisation issue comes as a result of a compromise between two traditions: the political aspect of the European project and the technical aspect of telecommunications. This ambivalence has been the basis of European policy, with a logic of inversion (harmonisation as the political facet and liberalisation as the technical facet).

In fact, there is a false symmetry in this: liberalisation is a dynamic concept, harmonisation is static. The former is simple and quasi immediate (taking a legal decision is all that is required); the latter is more complex and slow to actually put into practice (the condition for harmonisation to take place is of the 'all other things being equal' kind). In certain respects, harmonisation remains the trigger to liberalisation policies and this false symmetry has been the driving force behind regulation policies in the European telecommunications sector.

2.2 European integrated services digital networks: a formal industrial policy in the search for efficiency

The concept of harmonisation has been partly applied in the course of the definition of a European industrial policy, R&D programmes and standardisation policies. An example of standardisation as an industrial factor is demonstrated by European integrated digital services networks (Euro-ISDN) policy, largely inspired by Community action (as opposed to GSM where industry took the lead): support for the memorandum of understanding agreement by 26 operators belonging to the Confederation of European Posts and Telecommunications and the management of the User-ISDN Forum.

In June 1992, the Council of Ministers stressed the importance of giving ISDN a trans-European dimension. Euro-ISDN policy is of top priority in terms of the incentive policies advocated by the Commission. It certainly has the most highly-developed formal framework, the broad lines of which are of particular interest.

Identification of ISDN as a sector in need of Union direction is the result of a series of decisions taken by the EU.

In 1986, the Council decided that there should be a coordinated introduction of ISDN into the Community, in particular by implementing recommendations on standardisation. The Community was to put is financial means at the disposal of the investment required for ISDN.

In 1989, action was intensified and the decision was taken to create a pan-European ISDN system by 1992, based on common specifications taking account of worldwide standardisation.

In 1990, the Commission confirmed that pan-European infrastructures did not yet exist in this domain and that information on the real market was poor.

In 1992, the Council defined Euro-ISDN as being 'the implementation of ISDN fully-based on harmonised European standards conforming to the Memorandum of the ISDN agreement'.

Therefore, it considers the Euro-ISDN standards worked out by the European Telecommunications Standards Institute to be of the utmost importance. Network operators are being pushed to develop plans for the transition from ISDN to Euro-ISDN and to consult, in conjunction with the Commission, both users and manufacturers for an update on progress; member states must encourage the introduction of ISDN by taking account of the ONP; the Commission is to intensify negotiations, define requirements within the framework of the European Users Form, outline the means of communication for small and medium-sized enterprises (SMEs), favour the creation of a European terminals market and examine the possibilities of providing support specifically for the introduction of Euro-ISDN in less-advantaged regions.

In 1993, the Commission set out a series of detailed guidelines for the development of ISDN as a trans-European network. In doing so, it set in motion its action plan which was defined in 1990. Trans-European ISDN is considered to be a fundamental element of trans-European telecommunications networks. The objectives to be met include:

- the availability of a Euro-ISDN infrastructure, in particular a minimum range of services conforming perfectly to harmonised European standards;
- total geographical coverage by Euro-ISDN throughout all member states.

The following priorities have been set to enable these objectives to be met:

- to contribute to a rapid provision of Euro-ISDN for all professional users, notably SMEs;
- to maximise Euro-ISDN use within the Union;
- to facilitate, thanks to Euro-ISDN, access to regions less equipped with modern communications networks.

The following actions should allow both objectives to be met and provide the basis for projects in the common interest:

- the elimination of bottlenecks encountered during the installation of Euro-ISDN;
- a guarantee of full inter-operability for telematic services;
- the transition of public and private-sector applications to Euro-ISDN;
- the promotion of Euro-ISDN terminals.

Furthermore, the Commission has proposed that the Council adopt Union action on TEN-ISDN over the period of several years, in order to provide a back-up to projects of common interest and ensure total inter-changeability and inter-operability of ISDN applications. It has, therefore, determined three categories of action:

- projects on infrastructure;
- projects on transborder telematic services and applications: projects guaranteeing the promotion of ISDN use, including promotion of the availability of ISDN terminal-equipment;
- measures on end-to-end compatibility of a limited range of basic telematic services and corresponding terminal equipment.

Projects in the common interest may be subject to feasibility studies and support for their implementation.

It is somewhat difficult to be objective about the effects of a policy that has perhaps permitted the advance of a doctrine, just because it resulted in its implementation. Nevertheless, the end results are considerably short of those which might have been expected, given the stakes that ISDN introduction represent for European telecommunications as a whole.

2.3. RACE, a second instrument for a weak industrial policy

Expounded in 1986–87, to a slightly greater extent than ISDN policy, the RACE programme followed the identical approach to that established for the ESPRIT programme: the concept of a pre-competitive research policy which was re-applied to the field of telecommunications.

The main objective of RACE remains the introduction of integrated broadband communication (IBC) within Europe by 1995, with a target of 50% penetration by 2005–2010. With only the deadline established, such an imprecise definition of objectives certainly leaves much room for manoeuvre.

The international stage, which serves as a backdrop to the programme, is both a threat (American and Japanese initiatives on information and consumer electronics technologies) and a reason for hope (contribution to the emergence of a global economy served by open systems through international collaboration).

Assertion of the pre-competitive nature of the operation rests on:

- bringing the different economies of Europe up to one level;
- equal opportunities for European companies;
- the coherence and correspondence of decisions taken by different players, with regard to future European and world markets;
- re-asserting the pre-normative nature of the operation, a notion that reinforces pre-competitivity in addition to paving the way for standardisation.

Therefore, RACE seems to have illustrated more concrete political direction. But this may only really be brought about through the definition of the means by which this is to be achieved, which is evident for any research and development programme. RACE has become one of the structural elements of European telecommunications policy and thus a form (slightly weak as it is) of industrial policy.

Trans-European networks provide new scope for the deployment of ACTS (the former RACE programme). More accent is to be placed on applications, but the basic philosophy remains unchanged.

There is a risk, at least in its application, that some of the difficulties already outlined may be encountered:

- the notion of interconnection is only relevant if driven by consumers with a real need to be connected to each other, and it seems to be difficult to set up closed-user groups on a topographical basis (islands served by a trunk);
- a service lies between the application and the network. This cannot be overlooked.

Table 1. Trans-European networks-integrated broadband communication in the shadows of RACE

Section XII of the Maastricht Treaty has given legitimacy to this type of initiative, reinforced by the White Paper published by the Commission: 'In this document the new information technology was identified as a key challenge and an opportunity for European citizens to benefit from tomorrow's innovation...the TEN-IBC projects that are now working form and essential part of this strategy.' The applications retained as prototypes for wideband trans-European applications suffer from several handicaps: their size; the search for a forced multilaterality; the lack of long-term commitment on behalf of partners; their distance from the operators; the opportunism of consortia; an over-short time period; and an as yet undefined follow up of progress. The development of TEN-IBC has been fairly ridiculous: 14 dissipated projects; lack of thorough checks; absence of validation criteria; terms and conditions set out by people who will never actually pay for nor use the service; lack of any real commitment by consortia members.

The definition of the service poses three types of questions on: technical interfaces, the operator and the price. It is not certain whether European players are mature enough to take on this change in direction in the next two years.

It will be left to future projects, centred on techno-economic aspects, to establish the link between the strategic orientation of ACTS (for the provision of wide-band services in Europe) and its technical orchestration (for the definition of flexible services).

3. The European commitment to the information highway and its relationship with the US initiative

Until recently, American policy has not been a strong nor explicit enough reference to influence the determination of European policies: neither RACE nor Euro-ISDN have an equivalent in the US. Only the deregulatory wave originating from America has left its mark, but on closer inspection there are in fact very few similarities between the deregulation taking place in the US and the European model – that is until now. Yet, at the same time, it is emphasised that Europe seems to be flagging in comparison with the strikingly forceful trend in the USA today – takeover bids, mega-alliances, growing numbers of tests on pilot sites and announcements of record investments. All this presented under the aegis of superhighways...

Discussions among members of the European Union seem to be tightening around the neo-Keynesian aspect of the project, even though public funds have been bled white. It will be seen, however, that additional public investments in the United States have hitherto been strictly limited. On the other hand, three factors appear to be playing a mainspring role here. Firstly, a better climate for technological investments with lower interest rates and, more importantly, a far more positive reasoning with regard to the prospects offered by the information society. Secondly, a fantastic amount of bargaining on the part of the various players to break up the regulatory framework inherited from the Modified Final Judgement of 1984: the RBOCs wish to go back on the ban on using cable networks in order to diversify and profit from growth rates and a comfortable cash-flow, while looking at the possibility of merging with or taking over cable operators as a way into the territories belonging to other telephone companies; long-distance operators want to speed up the disintegration of the monopolies held by local telephone companies through the medium of mobiles, alternative metropolitan networks or cable. It is thus a strengthening of competition, including that resulting from the formation of alliances, which influences investments in superhighways. Far from designing a homogeneous network, this redistribution of roles provides the outline of a 'network of networks' in which interconnection is of paramount importance. Thirdly and lastly, emphasis is to be put on the extensive use of on-site testing to obtain a concrete evaluation of hypotheses on new markets with the associated technical configurations and partnerships.

3.1. The White Paper

Inspired by President Delors, the White Paper published by the Commission puts telecommunications and trans-European networks on a priority equally as high as that of the Clinton administration: a global policy has replaced a sectorally-oriented one, which supposedly favours the attainment of objectives in terms of productivity, growth and employment.

On the status of information technologies, the White Paper maintains that they are the basis for the changes currently taking place in our societies and the lever for future policies – at once both the cause and the remedy.

As a result of social changes, information technologies affect consumer behaviour and citizenship practices. It is the kind of opinion that may also be found in NII policy. But at no time is any precise evaluation made of qualitative and quantitative impacts. Neither the citizen nor the consumer are put to the test by this argument: the White Paper contains no analysis of social or financial demand. Therefore, one may question the method used to arrive at the level of investment required to face the challenges and the determination of priorities established to divide up this investment.

In the White Paper, the Commission estimated that the investment necessary for such development over the next ten years would amount to 150 billion ECUs. Priority projects are to require 67 billion ECUs during the period 1994–1999. 'Finance must be raised by private investors with national and community public authorities only intervening in a marginal and incentive manner.'

Within this perspective, the question of profitability cannot be avoided: absolute profitability (the pay-back of such operations); and relative profitability (the opportunity for investment with the aim of developing the market compared with investment aimed at controlling it).

Finally, one could question the method that would allow 'the will and means of a public-private partnership to be mobilised', leaving open the question of the subsidiarity of such a policy.

3.2. The Bangemann Group's mission

With liberalist lines (market forces are ready to take on and finance the arrival of an information society, on condition that the regulatory environment is as open as possible), the report advocates an incentive policy without, however, providing any answers to the questions implicitly brought up in the Commission's White Paper.

Mixing classical arguments with industrial policy, aiming to fight the activism of North American and Japanese industry and fully-backing accelerated deregulation, the report concluded with the necessity of strengthening Europe. The single market has proved inept at analysing current situations (no detail on the disparities within Europe, no stimulation of potential and financial needs) and lacks daring in its proposals (telework and research networks were already established in the White Paper; electronic highways are being treated as something of a gimmick).

But above all, there is in our opinion some difference in perspective between a policy specific to Europe and that which is advocated by the American NII. Comparative analysis is however of great interest, as Table 2 shows.

Table 2. Bangemann Report vision compared with the Clinton-Gore initiative

EII	NII
Diagnosis	
The global nature of the information society.	The competitiveness of the American economy is dependent on multimedia deployment. NII intends to be the driving force of technological development, a national 'market place' and a system of government.
Rationale	
Reactive to the US initiative.	A key tool for a long-term strategy of economic development. Political intentions: the Democratic administration wants to gain the support of the high-tech industry.
Context of the industry's regulation	
Deregulation.	Oligopolies formation.
Financing	
Private in its principle; redirection of public funds with respect to former goals.	Private in principle; backed by public funds.
Evaluation of the demand and requirements	
Identification of ten lines of application, retained for their illustrative effect.	Drawn mainly from the demand of companies who expect the NII to bring public networks up to standards.
Expected players rationale	Mobilised by the effects of competition.
Principles	
Acceleration of the liberalisation process.	NII will develop through competition; FCC will resolve difficult competition issues.
Priority accorded to network interconnection and the inter-operability of services and applications.	Reflection about the tariffing of services (using the Internet model).
Reinforcement of guarantees on intellectual property and data protection.	Political preoccupation: stopping the growth of social inequalities (maintenance of universal service).
Standardisation and spreading of trans-European basic services.	Encouragement of private investment.
Public policy	
Negotiation of reciprocity agreements.	Promotion of applications which create a learning effect.
Support given to Euro-ISDN.	Promotion of the Internet on the international market.
Support for the development of a broadband infrastructure connected to cable and satellite.	Promotion of actions aiming to network federal agencies and develop electronic information exchanges.
Increase awareness among SMEs, administrations and the young generation.	
Strengthening of competition policies.	Proposition of laws.

4. Issues pending on the European information infrastructure policy

Public policies are currently looking for both a doctrine and the means of intervention which avoid calling into question the principles on which 15 years of European policy in the telecommunications field have been defined: progressive deregulation therefore rendered unavoidable, and a weak industrial policy.

The 1987 Green Paper constituted a document which designed a doctrine based on European consensus, that of the lowest common multiple. The harmonisation-liberalisation diptych is the result of a compromise involving two traditional factors: the political aspect of the European project and the technical aspect of telecommunications. With its inversion arguments (harmonisation as a political factor and liberalisation as a technical factor), this ambivalence lies at the base of European policies.

Table 3. The trans-European broadband backbone (TBB): an early concept for European superhighways

A – Concept

'The TBB is defined as the concept for a unified approach by the telecommunications operators to provide a main support for the early introduction of advanced telecommunications services by providing high-speed digital capabilities beyond national borders in the European Community.

'The TBB is the common approach for a European-wide network of digital telecommunication "highways" across the borders within the Community based on the existing and planned capacity for current services and enhanced by a capacity to support the early introduction of advanced telecommunication services.'

B – Services and networks

'The services taken into account for the TBB project are all kinds of digital or digitised services originating from the national networks.

'There is a need for a TBB to provide in advance of 1995 the appropriate transnational infrastructure by using available technologies to render possible the Community-wide introduction of advanced telecommunications services.

'Previous considerations lead to a first-reference switching-system architecture for the TBB comprising the following dedicated module:

- 64 kbit/s ISDN switching module being the digital part of the existing international gateways.
- Broadband switching modules composed of space division matrices, with individual control units. Each module can be dedicated to one defined bit rate. The first bit rates to be taken into account are 2 Mbit/s and 140 Mbit/s.

'The control units of the modules should be interconnected via a control information network which will initially be based on national prebooking centres and signalling procedures according to CCITT No 7.

'It is recommended to standardise with high priority the basic broadband user access, in particular the basic user access interface and the user access parts of the signalling procedures (including those appropriate for prebooking centres).

'Considering the general characteristics, capabilities, advantages, and disadvantages of present and future satellite systems, it is recommended to use these systems for the early implementation of the TBB to peripheral regions, as a backup to the capacity, for special call configurations (broadcast), and for the early linking to subscribers.'

C – Policy and players

'There is a need to have a TBB as an integrated part of and embedded in the adopted programme of actions of the EEC in the field of telecommunications to support the international competitivity of the European economies in general and of the telecommunications industry in particular.

'The implementation of the TBB-concept needs a set of collaborative arrangements for the various actions concerning research, standardisation, planning, implementation, marketing and finance:

- a common agreement on the introduction strategy;
- a commitment to the chosen aims by the member states and the telecommunications operators as appropriate;
- a number of common facilities provided by the telecommunications operators and – in some cases – with Community participation;
- a number of joint actions/ventures involving the telecommunications operators.'

Source: ETCO, The trans-European broadband network. Report to the European Commission, 1985.

This, in fact, is a case of false symmetry: liberalisation is a dynamic concept, while harmonisation is a static concept; the former is simple and virtually immediate (all that is required is a legal decision), the latter is complex and slow to implement (the condition for harmonisation is of the 'all other things being equal' type). In some respects, harmonisation is the exploitation of liberalisation policies, and this false symmetry the mainspring of European telecommunications regulation policies.

Changes in the US administration have brought about a different prospect: with a mix of Keynesianism and consumerism, the US economic programme has placed telecommunications at the heart of the national policy system for the first time. The Europeans, who are otherwise busy with post-Maarstricht gestation problems, are consequently going to endeavour to allay suspicion by changing the colours of their already-old policy of trans-European networks for today's – those of the information superhighway.

Two important events must be taken into account in this context: the publication of the Commission's White Paper and that of the initial conclusions of the Bangemann Report. Under the inspiration of Commission President Jacques Delors – macro-economic concepts and political pedagogy – the White Paper puts telecommunications and trans-European networks on the same level as that decided by the Clinton-Gore administration: it is no longer a question of a sector-based policy but of a transversal policy which is supposed to accompany or favour employment and productivity-increase objectives.

Prepared rather hastily, the telecommunications section of TEN policy as it appears in the White Paper presents more questions than solutions. With its liberal inspiration ('market forces are ready to take responsibility for and finance the creation of an information society' provided that the regulatory environment is as open as possible), the Bangemann Report advocates an all-embracing incentive policy but without supplying answers to the questions implicitly raised in the White Paper. Mixing classical reasoning with industrial policy, with the intention of countering the activism of Japanese and North American industry and pleading in favour of accelerated deregulation, the report declares the need to consolidate the European economy – the great market – without having given real proof of precision in analysing the various situations (no trace of European disparity, no assessment of potential and fulfilable needs), nor of any real boldness in its proposals (telework and research networks were already in the White paper; highway teleguiding is regarded as gadgetry).

Until now, all accent has been placed on the political dimension of the problem (development of the supply of trans-European networks in order to adhere to common market objectives and meet international challenges), without giving much attention to the evaluation of demand and the determination of its rationale.

In viewing current history, in which contingency and inertia factors try to find expression, it is necessary to recall 1987 when in formalising ten years of telecommunications policy in the Green Paper, the Commission had included *in extremis* the ONP concept which was supposed to be the tactical reply to developments in progress in the United States. This comparison leads us to bring up two points, two questions that seem relevant within the framework of this contribution:

1) Public policies often have to administer a tactic (the most urgent) and a doctrine, but do they strive to clearly represent the interests of both aspects?
2) Since the publication of the Green Paper on telecommunications, the debate has been substantially enlarged; today, telecommunications has a status all of its own in regard to the economy in general and thus to the construction of Europe in particular. What role should this play in the actual formulation of such policies?

On the status of information technology, the White Paper supports the view that it is the foundation of the changes which our societies are undergoing and, at the same time, the lever to be handled in formulating policy, the cause and the remedy.

Table 4. An evaluation of the demand for trans-European networks

We estimated that by the year 2000, demand for trans-European communications will amount to twelve billion ECUs. This is a modest figure: it represents just 8% of the European telecommunications market, itself estimated to be worth 147 billion ECUs by the turn of the century.

This figure could even be considered 'absurd' upon the basis of two references: firstly, a simple comparison of macro-economic data (European exchanges/GDP of the twelve Union member states), which puts the market at 30 billion ECUs; a second comparison with the air-transport sector, which certainly has its limitations, may also be of use. Here, the relationship between the market held by European companies and the trans-European turnover (freight and passengers) is also highly significant, as it could be in the order of 25% by the year 2000.

And yet, in our opinion the demand for trans-European communications is not in the least bit under-estimated: analysis of the dynamics of both micro and generic markets shows that current inertia based on objective constraints and subjective forecasts leaves no prospect whatsoever for a sudden increase in demand in the medium term (five or six years).

Voice services will make up 70% of this market, an important fact which it would advisable not to overlook: by the year 2000, universal service (essentially vocal) will therefore remain the basis of trans-European communications. Trans-European infrastructure networks, according to their figures, could be worth four billion ECUs, that is to say half the turnover made by the universal service.

We could then envisage two scenarios considering both the level of demand for trans-European networks and the policy instruments already in place or starting now:

- scenario one: it may be considered that the market estimate is in line with the dynamics of European construction; it is therefore a question of following current developments, essentially by means of a policy to manage the bottlenecks, the corner stones of which would be regulatory policy;
- scenario two: it may be judged necessary to engage a comprehensive policy to stimulate demand so as to raise it to a level that may be expected of an integrated European economy, that incorporates a wide and intensive use of telecommunications; it is thus an entirely new industrial policy that must be implemented, one that is more oriented towards providing support.

Scenario one poses no particular technical problem: given moderate demand, providing a valid interest in regulatory developments is enough (while respecting established terms) so as not to interfere with the few initiatives in progress and ensure efficiency in the application of ONP policies.

Scenario two is or more interest in terms of the options available for political action. It implies that the search for realism in the definition of a demand-stimulating policy be laid down as a rule. The first concern must be to get this policy in motion. A chart going from the most urgent to the most important is an interesting and pragmatic way of handling priorities.

Source: IDATE, research under contract with the French national regulatory authority, the DGPT.

In advance of social change, information technology has an effect on consumption norms, and subsequently on the exercise of citizenship. It is a case of logorrhea which is similarly found in NII policy, but at no time is the quantitative or qualitative impact described. Neither the citizen nor the consumer is 'put to the test' in this reasoning: in the White Paper, there is no analysis of the social demand nor of the fulfilable demand.

One could ask about the method that was used to fix the level of investment required for the deployment of the EII. Another issue relates to the return on investment that can actually be expected for private investors. Neither is it clear by which method policy-makers will be able 'to mobilise the willingness and the means on the basis of a public-private partnership'.

In all cases, the EII policy will suffer from a 'double-bind' which is usual in the European context. On the one hand, consensus will be reached on the minimum base; on the other hand, policies will have to test the limits of such a consensus, making explicit internal antagonisms.

II Alliance or Competition: Company Strategies at the National and International Level

Telecommunication:
New Dynamics and Driving Forces
A. Bianchi and G. Richeri (Eds.)
IOS Press, 1996

Chapter 3

Dynamic industry structure as *policy*? three alliances and world telecommunications

David Allen
Concord, Massachusetts, USA

Abstract: In one year, the stage for global telecommunications has been set. Three world alliances have emerged, from combinations initiated by four of the five large telecommunications companies. This chapter asks what global industry structure do we want, and get, beginning with the alliances. The analysis proceeds from an unconventional view that centres on the technical, and social, evolution in networks. To encompass the swings between innovation and standardisation, a 'community of competitors' gives twin-policy billing to competition and consensus. An industry structure to suit, necessarily iterates through a dynamic routine – alternately fragments, then monolith – as a matter of conscious policy.

1. Introduction

It is less than two years since BT in the UK, together with MCI in the US, announced they would join forces to supply global telecommunications services. As expected, that announcement triggered a headlong rush. The other major world telecommunications companies moved vigorously – they intend not to be left off the new, global stage. The developments have been rapid-fire and, in some cases, even surprising.

Deutsche Telekom and France Télécom began to pump life into their Eunetcom joint venture, to consolidate their position in Europe. Subsequently, they have announced an alliance with the *other* 'second force' carrier in the US, Sprint. Ostensibly, this new configuration is the global entry from (continental) Europe. AT&T has announced – they did so when BT-MCI went public – its WorldPartners alliance. The company has 'not looked back' as it aggressively has built a number of linkages around the world, including one with Unisource for Europe.

After just one brief year, the stage for global telecommunications was set, with four of the five large telecommunications companies – AT&T, BT, Deutsche Telekom, and France Télécom (in alphabetical order; NTT, in Asia, will appear below) – striving forward toward a new, global regime. The result is three aspiring entrants for the global arena, each with a dominant partner(s) from a strategic region of the industrialised globe [1]:

	dominant 'centre' in:
• BT-MCI's Concert	UK
• Deutsche Telekom-France Télécom-Sprint [2]	Europe (continental)
• AT&T's WorldPartners, including Unisource	US

Asia, the third leg of the triad, is conspicuous by its absence from the right hand column.

What industry structure, global and otherwise, do we want for telecommunications and the related technologies with which it increasingly converges? What industry structure are we likely to get? Recently I have suggested that we can answer the structure questions only if we are prepared to incorporate the dynamics also. This view is unconventional overall, and for industry structure it prescribes iteration through a dynamic routine as a matter of conscious policy.

This chapter addresses the normative and the descriptive questions for global telecommunications industry structure, when taken from this novel view. The three main entrants for global telecommunications are assessed from the new perspective, for what-do-we-want and what-do-we-get. But at the same time, this new conception is trialed against the complex real-world case. Surely such mutual testing, where the idea and the empirics reflect mutually upon each other, is our process-of-choice to evolve more useful conceptual tools.

We begin with preliminaries – we establish salient details of the three new alliances, and we also briefly review the new conception for industry structure and dynamics. With that as foundation, we look again at the new entrants to global telecommunications, now with the new view. Then we can turn to what we want and what we get, structurally. Dynamics as policy is a central focus.

You are invited to test both the new idea and your notions about the global structure which is emerging, when seen through the new eyes.

2. Preliminaries

2.1. The three new alliances

What are the particulars for these three telecommunications alliances, now setting forth on the global stage?

For our purposes here, the first two are straightforward. BT-MCI's Concert and Deutsche Telekom-France Télécom-Sprint are alliances among two or three members, with each alliance intent upon becoming a (if not the) global service provider to multi-national companies, the customer group which is perceived as the vanguard for globalisation. For BT-MCI though, there is an added fillip. Concert has now established several 'distribution arrangements,' that is, it has engaged local sales agents. In Japan, the sales agent is Nippon Information and Communication/NI+C International – NI+C International is a joint venture of IBM Japan and NTT [3].

The composition of AT&T's WorldPartners is another story. Initially, partners divided between equity and associate status; that has changed [4]. Now all partners are members, while some are also owners. They are sprinkled around the globe. At mid-February 1995 the WorldPartners were:

In Asia:
member
- the Philippines: Philippines Long Distance Telephone Company
- Hong Kong: Hong Kong Telecom International
- South Korea: Korea Telecom
- New Zealand: Telecom New Zealand International
- Australia: Telstra

member and owner
- Singapore: Singapore Telecom
- Japan: KDD

In the Americas:
member
- Canada: Unitel

member and owner
- the US: AT&T

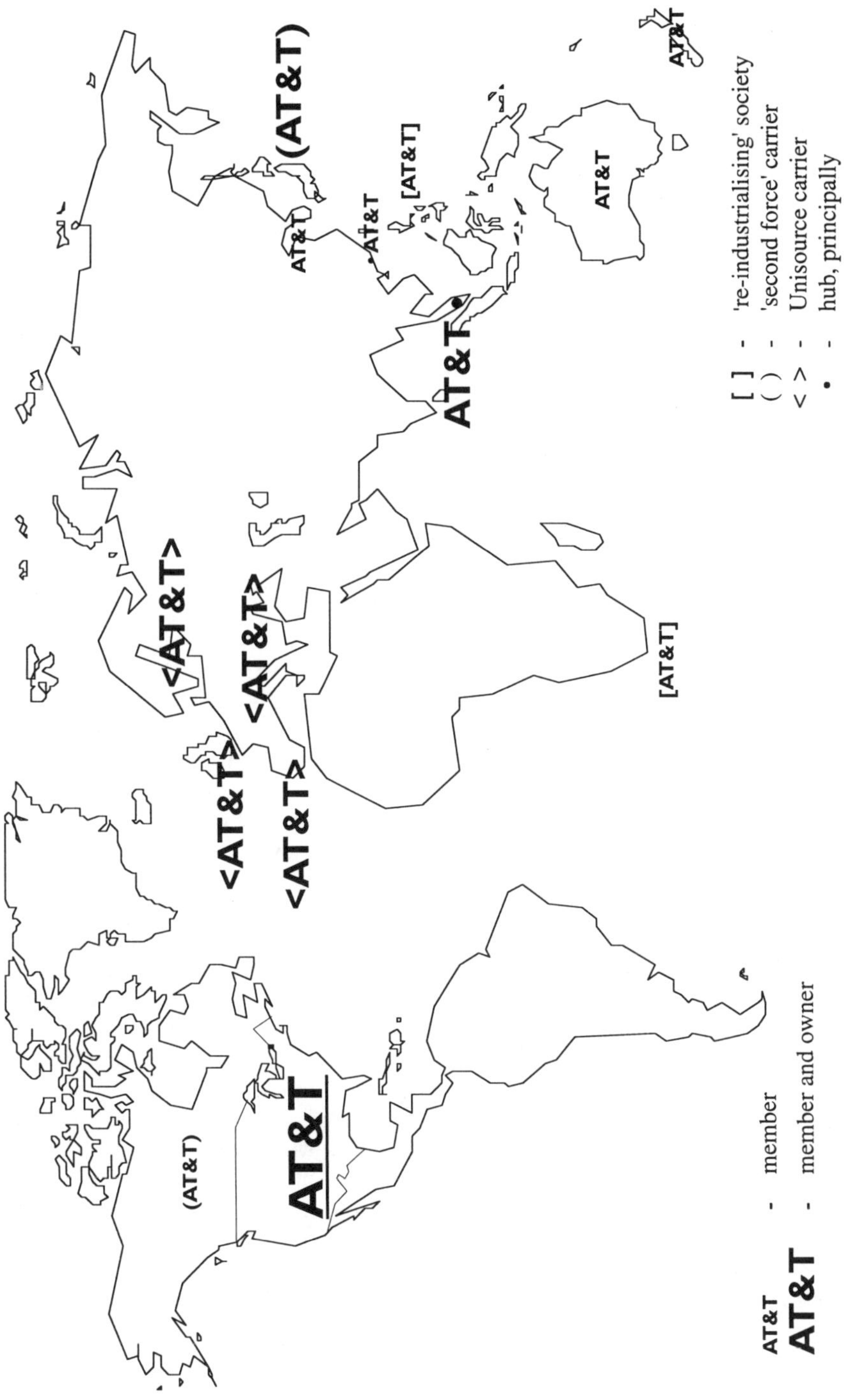

Figure 1. AT&T's WorldPartners: by internal structure

In Europe:
member and owner
Unisource [5]

- Sweden: Telia
- the Netherlands: KPN (new name for PTT Telecom of the Netherlands)
- Switzerland: Swiss PTT Telecom
- Spain: Telefonica

(The Unisource relationship with SITA is not part of WorldPartners.)
In Africa:
member (announced) [6]

- South Africa: Telecom South Africa

Figure 1 depicts AT&T's WorldPartners geographically. Again, there is an additional fillip. In Japan, WorldPartner KDD has agreed with NTT to interconnect their virtual private networks [7].

2.2. *Telecommunications industry structure, in the new view – a brief recap*

The new conclusions about industry structure grow from an underlying logic. We first summarise the position on industry structure, then review the underlying argument [8].

2.2.1. *Industry structure*

To accommodate the dynamics, industry structure is conceived as 'nested.' For instance, the European Telecommunications Standards Institute (ETSI) in Europe is a nesting of the region's carriers (and others) [9]. The key feature of the nest is that individual units can act independently, when called for, but the units can also come together as a larger group, when concerted action is indicated. Hence, the structural dynamics: alternate separation into fragments, then coalition into a temporary monolith, and repeatedly.

Policy dictates the points of inflection between industry-as-parts and the industry-as-whole. To use more direct words: conscious, joint choice selects the occasions which require a joint approach, and it also plays a role in the reverse. ETSI standard-setting, for instance, represents a joint choice that the time has come to settle on some technological suite, an agreement which is essential for a network to interconnect and operate. Equally, innovations are pursued more locally, independently, company-by-company, new ideas which later will feed some more collective standard-setting procedure.

Both part and whole have their place. The idea of virtual hierarchy is now used occasionally [10]; in the conception here, it would take on pinpoint meaning – the industry group operates as *both* hierarchy and peers. Also, the ETSI example makes clear there is no clearcut temporal sequencing, but a much more complex situation in which separation and coalition likely occur in parallel. For instance, when the same company is active in different technologies, it will probably find itself both independent and grouped at the *same* time. Hierarchy relations and peer relations are indeed intermixed.

Is this Coase and Williamson simply warmed over (or standard microeconomics on vertical integration generally), or management theory on centralisation/decentralisation? Decidedly not. Here there are frequent, intermixed shifts, back-and-forth between grouping at some industry level versus acting as an individual company. Those shifts are a matter of conscious, promulgated policy.

Some of the structural implications: power relations, not surprisingly, are central. If one or a few companies simply dominate the industry, lesser brethren are not likely to accept a coalition, certainly not the fluid repeat changes which are necessary – but some companies will surely be more equal than others. Power relations, as a result, here take a form that acknowledges leadership but which, at the same time, confers peer status on the smaller member. Also, to sustain the global and the local equally, as the virtual hierarchy requires, puts an emphasis on two basic capabilities. One is clear definition of the company and its boundary, while the other is the ability to link well with others when required.

2.2.2. The underlying logic

These conclusions about structure are driven by the underlying requirements for continuing innovation in a network. Repeated cycles between innovation and standardisation demand an alternation between competition and consensus. When the time arises for a set of competing new ideas to be melded into an interconnectable standard, joint choice is necessary to move from competition to consensus. That is the structural point of inflection between industry-as-parts and the industry-as-whole. (Technically, that is a point of critical mass, and the reverse is Schumpeter's 'creative destruction'.)

This turns upon the human capacity to shift viewpoint fluidly among levels of the nested actors, depending on the issue in hand – when innovation is the question, the focus is more private; when standardisation, more public. One key effect is to maintain local variety, within standards agreed by a larger community. The policy regime which is implicit here might be called a 'community of competitors'.

Market evolution depends upon the dynamic nested structure operating within this policy regime. Competition and consensus iterate repeatedly and, as we have seen already in structure, they intermix inextricably. The competition is principally among ideas and takes a plus-sum form, to permit the dynamic shifts. Competition is vital, but so is its opposite, consensus.

2.2.3. Conclusion to the preliminaries

Some might ask, is it reasonable to suggest that such a complex arrangement be attempted. I believe, instead, the question is whether any *less* complex conceptualisation reasonably captures the reality we live with daily. If we need an illustration, the intellectual property rights/IPR policy which ETSI intended (and which was stymied by American objections) serves precisely [11]. One way to express the essence of the intended ETSI policy is that companies bring their individual innovations to the table, to be incorporated into a vehicle which will serve a wider public. This of course is the point of inflection described above, and the shift from the individual level to a higher level in the nest.

On the supply side of the market, which is our concern here, this regime dictates a culture of the industry. The group requires social coherence adequate to see the members through the repeated swings away from each other and back again. In the parlance, this is a community of the industry, with tacit and evolving but mutually accepted rules of the social road. Economics [12] has long grappled with the challenge of trying to understand oligopoly. Perhaps we begin to unearth here where some of the complexity lies, and why we may reasonably expect the understanding to be a challenge.

In this new view, telecommunications industry structure, our direct focus, makes sense when seen dynamically. But the structure is dynamic in a particular sense. The dynamics are conscious, as a matter of policy, not the fickle product of some 'unseen hand'. They are iterated. And they are intertwined, as described above.

Now we turn to use these ideas, both in their normative and in their positive forms. We begin by briefly looking back at the three global entrants, framing them in the new light.

3. The three global alliances – in the new view

How do the three alliances appear, when set against this new perspective?

3.1. Deutsche Telekom-France Télécom-Sprint

Deutsche Telekom has articulated a vision for Europe which, when taken globally, is in synchrony with a community of world competitors[13]. This vision proposes, among others, a single group of carriers where relations within the group are consciously a mix of competition and cooperation[14]. Whether this vision will translate to the new alliance among the three companies remains to be seen, but it is the path required by the new view summarised above.

The Deutsche Telekom position pretty clearly calls for a transition from the relatively casual correspondent relations which are presently the style for the industry. The new mode envisioned amounts instead to a social contract, which would effectively create a community of the companies. Looking again to Europe for an example, we can ask whether any European company may claim to have adopted this new style *internally,* within itself. But the European Union approach to leadership – where the head position in the hierarchy is handed around to each member of the group successively – is a vital element in the sort of power relations which are necessary[15].

3.2. AT&T's WorldPartners

AT&T has assembled a large array of partners, a group that is wider and more geographically dispersed, by a good deal, than the others. Is this too a new community?

We need to ask, what are the power relations among the eleven who make up the WorldPartners group. AT&T's strength and market position stand out – it towers among this particular group. Except for three of the partners (and AT&T itself), the participants are carriers for a 're-industrialising' society; 'second force' carriers, against already established incumbents; a new entrant; or principally hubs[16].

Carriers for a 're-industrialising' society:

- the Philippines: Philippines Long Distance Telephone Company
- South Africa: Telecom South Africa

'Second force' carriers:

- Japan: KDD
- Canada: Unitel

New entrant:

- Europe: Unisource

Hubs, principally:

- Singapore: Singapore Telecom
- Hong Kong: Hong Kong Telecom International

Telstra in Australia, and Telecom New Zealand International in New Zealand, are two partners with the market power which ex-PTT status, and universality, confers; and their base is an industrialised society. For some sense of scale in these two cases, the 17 million population of the larger (Australia) is roughly five per cent of Europe, seven per cent of the US, or 14 per cent of Japan – although Australia's position in world affairs is a good deal larger than the relative size of its populace.

Korea Telecom serves one of the veteran Asian tigers, and with a population about two and a half times that of Australia. South Korea would compete with Japan; in selected market niches, it does so effectively.

These appear to be the three strongest WorldPartner market positions, besides AT&T.

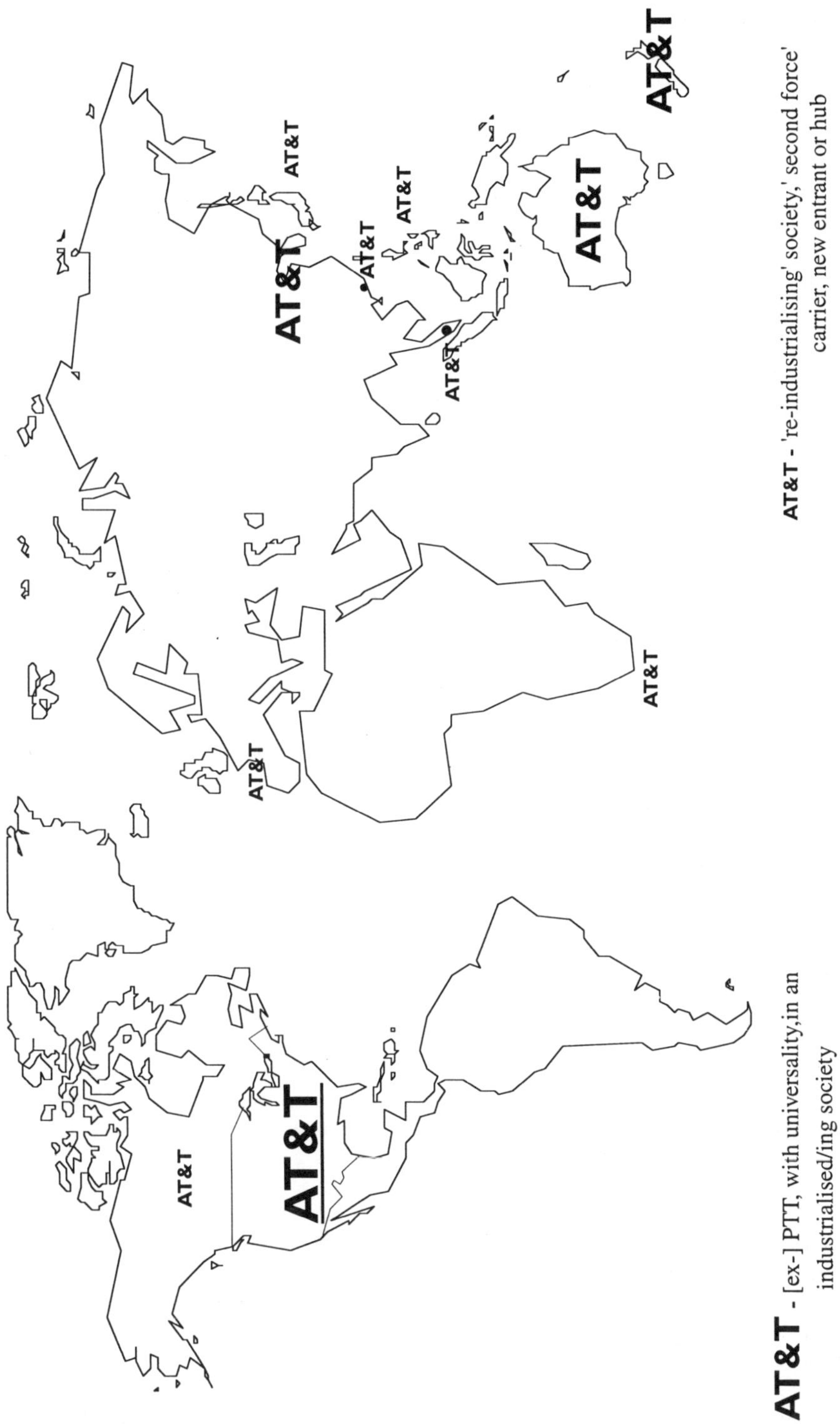

Figure 2. AT&T's WorldPartners: by market force

None of this gainsays the WorldPartners position worldwide, of course – AT&T's apparently commanding lead globally is our focus below. But AT&T towers over the group, without peer. Figure 2 considers AT&T's WorldPartners from the point of view of market force.

The question here is whether and what social milieu emerges for the partners. The question is particularly acute for WorldPartners, with eleven members (and there are 14 if Unisource is really four), since prevailing ethos becomes almost unavoidable with a sizable group. When seen from the point of view of its partners' individual market positions, as Figure 2 summarises, AT&T's role in the group could readily be interpreted as dominant.

On the other side, WorldPartners' stated position is the reverse, saying that no single viewpoint can dominate. Further evidence is a terminology which has been adopted inside the company, that of 'lead members' and 'support members.' Lead members are those in regions with the bulk of multi-national companies – that is, the regions with the WorldPartners' target market, but where the lead member likely needs support from the whole organisation. This new division between lead and support is the emergence of hierarchy which *might* differ from dominance.

3.3. *BT-MCI's Concert*

So far, this review (perhaps tacitly, but nonetheless) has inspected each group for the impact of the largest member. If we reverse the lens for Concert, MCI is the focus. Besides allying with BT which is the ex-PTT in the UK, MCI for other purposes has also partnered with the established incumbent in Canada, Stentor. Perhaps BT's approach to MCI was no surprise, but MCI's success with Canada's established carrier does snare our attention.

That MCI was the partner chosen by both BT and Stentor signals to me that the MCI ethos may play a role. The MCI position is a charming mix of the rebel's rhetoric from the days of McGowan's breakthrough into US telephony *and* the community-oriented solidity of the staid AT&T *pre*-divestiture. This MCI vision can be defined relative to Deutsche Telekom's. For MCI, the good regime is competition among *several* of the Deutsche Telekom sort of groups. Thus, relations within any *one* group are to be the familiar mix of competition and consensus – 'cooperatition', to use MCI's term [14].

The prevailing culture in both Commonwealth countries, the UK and Canada – although holding much in common with the US – nonetheless has more 'central tendency' than does the US, I am prepared to speculate. Such a central tendency is one seed from which forms of community might grow (so might also a totalitarian regime). Overall, the billing in these two countries may go more to competition, while community and consensus proceed more at an implicit level. A parallel between MCI's ethos and its allies' cultures is certainly no more than suggestive, but the fit *is* interesting. With the two Commonwealth countries somewhere to the left of the US regarding the individual and competition, MCI's position is somewhere to the right of Deutsche Telekom and a community of competitors [17].

3.4. *The three taken together*

We can draw the three points on this spectrum explicitly. When the three alliances are framed together, in this chapter's new view, BT-MCI's Concert appears intermediate between Deutsche Telekom-France Télécom-Sprint and AT&T's WorldPartners. Deutsche Telekom has effectively enunciated key elements of the vision implicit in the new view. AT&T, on the other side, appears potentially to be a dominant partner among a group that is now attaining some size, when dominance is anathema in the new view. (Although as already noted, there is evidence that WorldPartners may be moving toward a social structure which could be more consonant with the new view.) MCI (and BT ?) appear to be some mix of the two. Borrowing from a chart I drew in the earlier paper on vision, we can picture the three along a line (although without an adequate characterisation for the AT&T vision, nor complete information about BT-MCI – these questions demand further research).

Global Telecommunications:
Structure and Regime

Vision		
Alliance		
Community of competitors	*Competing groups, which are...*	*[? Competition]*
Deutsche Telekom-France Télécom-Sprint	[BT]-MCI's Concert	AT&T's WorldPartners

Figure 3.

But each of the three alliances has one quality in common – and it overrides the rest of the analysis. Each alliance implicitly has a structure of dominance itself, with one outsize, more powerful figure apparently in a position to command (in Europe, that is a pair). Put another way, these are *not* alliances among the major telecommunications companies themselves, as a group. Whatever public pronouncements might favour the new view, the alliances at present are not a community of competitors among the most powerful actors. Since an alliance among *that* group is the real subject of the new view, we cannot say that the process has really even started yet.

A final note: in the analysis so far, the carriers from Asia appear not to take lead roles – whether that is the case will be grist for our conclusion.

In this new view of the globalisation process, the lack of joint work *among* the major telecommunications companies stands out most starkly (France Télécom and Deutsche Telekom the exception, of course). That becomes the subject as we turn now to ask what we want.

4. What global industry structure do we want – the normative

The ideology of competition – the wave that has swept the world's telecommunications communities – is clear: the ideal is several companies in vigorous rivalry with each other. But a community of competitors would see those same rivals joined from time-to-time as one monolith. In this vision the major telecommunications companies must be capable of acting as one, as well as rivalrously, for a global network. How do we sort out this disagreement?

4.1. Nested supplier communities

Does the new view argue for dividing the network geographically? No. Interestingly, each of the three new alliances has, for their own internal purposes, done just that: divided the globe geographically. In general, one member does not operate on another member's turf when conducting the alliance's business [18]. So any of these models – if applied eventually as a global regime – *would* divide geographically. But particularly the history now accumulated with radio, notably cellular, makes clear that multiple operators are workable on the same terrain.

The question is not whether companies divide geographically, but whether the nested industry *communities* do so – whether the nesting is geographic. This is a challenging subject, worthy of some dedicated attention. Tentatively, the need for the supply side, the network, to reflect the peculiar varieties of usage from the demand-side – the geographically-based communities where we live and work – makes a powerful argument for geographic nesting in the industry. But the dramatic rise of electronic virtual communities – already with us over the ages *non*-electronically as diaspora, but now newly resurgent on the Internet – makes clear this is a complex question.

Such global nesting is a complex affair in its own right. At least three main levels seem likely, particularly if the new alliances are taken as some guide. The middle level calls for strong regional identity and organisation, a firm regional entity situated between the domestic and the global – a Europe, a North America, or an Asia, as the industrialised examples. Such an arrangement – what otherwise would be called regional blocs – simply sharpens the contrast between this conception and received views on, for instance, trade theory.

How, then, would the arrangement in this new view operate? There may be multiple providers. But their emphasis, rather than some commodity service, is more on producing desired local variety, within some interconnectable standard. The principal competitive rivalry – which too is crucial – revolves around new ideas, for on-going variety...which, when we think about it, may be a fairly reasonable representation of experience. It is here that the received and the new views start to touch each other.

4.2. Spoilers

Just as undue exercise of influence – 'monopoly', in the extreme case – spoils the competitive ideal in the received view, so in this new view there is also a spoiler. Here however, the iterated dynamics are spoiled by a *refusal to play according to community routines for* resolving *differences*. The potential spoiler loose among the three new global alliances is the US government, with its inclination to use the Deutsche Telekom-France Télécom-Sprint alliance as a bargaining chip.

US-government policy holds up the competitive ideal as the grail guiding, and blessing, its arm-twisting. But if the policy misfires, France Télécom and Deutsche Telekom may fail to field a viable global entrant [19]. That *could* hasten AT&T's accession to global dominance – and so become a marvellously ironic squelch to prospects for any US *global* ideal of competition.

(If dominance is not the spoiler in the new view – but dominance nonetheless disrupts community – how is dominance treated in the new conception? The conclusion, below, aims to clarify a bit.)

5. What global industry structure will we get – the positive

Is it really reasonable to imagine repeated swings back and forth, sometimes the industry tightly linked together, at other times individual companies proceeding quite independently?

5.1. The network firm

The industry structure – and dynamics – in Italy's Emiglia-Romagna region is perhaps one of the best-known 'real live' cases. Japan's *keiretsu* are also well-known (if not notorious, at least in classically liberal quarters). Virtual hierarchy – a network firm – takes on vibrant meaning in these places. The actors alternately assume either industry or company roles, depending on the occasion. Nor does technology guide the alternations. Technology in the form of electronic networks may support such organisation, but it is the human inclinations which surely take the first steps. (So, other societies with an interest probably do best to *start* with attention to their own cultural predispositions, before they then look to technology such as electronic data interchange.)

But if such highly-developed practice is showcased more than it is found in widespread use, what can we expect for the global network?

5.2. *Reflecting on the future*

While history is an inadequate reflection for the future, it can still be instructive (and really is all that we have). For the case of the global network future, consider two alternative paths from the past.

One is the history of AT&T's rise to the position of strongest carrier in the US (and indeed, on occasion, the world), a history that is now a century long but with salient features that are especially concentrated in a few decades just before and after 1900. At the time AT&T was in a fierce struggle with a number of quite scrappy carriers. But AT&T had special leverage in its hold on the US long-distance network. Although many factors would be pertinent, AT&T was particularly able to use its discretion over long-distance interconnection as a key lever to give it decisive advantage.

If we look back to Figure 2, AT&T has begun to assemble a far-flung set of *internationally* long-distance nodes. Although Singapore Telecom and Hong Kong Telecom International are 'only' hubs, they both will continue for some time to enjoy monopolies on their transit traffic. The parallels with the domestic rise of AT&T are seductive.

But the other history is Finland's. It is an unusual case, one seen in few other places, but therefore all the more instructive. In Finland, for reasons related in part to a Russian occupation, localities – and separate telephone companies to serve those individual localities – developed significant strength. A national PTT also developed, but to an important extent the PTT inherited the localities where extreme conditions most militated against affordable service. The PTT was responsible for long distance, as was typical for PTT's, but it was also responsible in the main for the more difficult, high-cost local service. The denser local traffic, in the larger cities and more hospitably-located regions, flowed instead through the independent local companies. The result in Finland – rather than the rise of one dominant actor – is a facing off of true relative equals: the local telecommunications companies as a group on one side and the PTT on the other. And when the two sides need to settle a problem in common *between* them, the sauna is always available.

Consider Europe again: Europe's telecommunications community process shares some aspects of the Finnish experience, with relative equals sorting out the questions together. If *this* takes hold as the global path, rather than dominance there could be the prospect for a community of relative peers, able then to undertake the more complex dynamics necessary for a fruitful future, globally, regionally and locally. In a similar vein, Unisource – although it is indeed an entrant in Europe – comes on board in some senses peer to everyone else there, since Europe is really without any *European* incumbent. That, once again, is the basis for a robust community [20].

Dominance or a community among peers – which do we get globally?

6. Power and community, the essentials – does Asia take a hand, too?

The choice is between a balance of power *or* the power of community protocols. A balance of power may tilt and allow dominance. Community protocols on the other hand (the particular sort enumerated here, that is) explicitly address the power relations, reserving peer space for those who might otherwise be dominated [21]. Between them, these two alternatives bracket the essential contrast between the two approaches.

We may not yet be able to say which of the two we get globally, but we can notice where the latter – some would say novel – community approach is in fact being used.

6.1. Europe

European examples have figured regularly in this chapter to illustrate the community view. The intended ETSI intellectual property rights policy particularly pinpointed the community dynamic. More generally, Europe has for some time pursued a dual policy – competition, to lower rates and improve generally, and also consensus, the opposite of competition, to bring harmonisation across a crazy quilt of national protocols. This is the tensioned regime prescribed in the new view.

The European policy discussion typically has, we quickly note, *not* acknowledged the tension between the two halves of the policy. But that too is beginning to change. See, for instance, a recent paper from Oliver Stehman, Commission of the European Communities, DG I [22]. The paper's discussion is concerned in part with European competition versus integration.

Indeed, the presentations of two papers which opened the 9th Euro CPR conference suggest that awareness of the basic tension may be accelerating in Europe. Jean-Claude Burgelman concluded by pointing to 'two rival policies as the basis in Europe: one central, for infrastructure; the other de-central, for subsidiarity' (see Chapter 1). Didier Pouillot, presenting for Jacques Arlandis, spoke even more directly to the point. He specifically contrasted policy for harmonisation with policy for liberalisation (see Chapter 2).

6.2. The US

Those developing the Internet in the US have evolved a 'gift' culture, where hackers contribute their work toward a better network for everyone. The innovations are then taken into a process which rapidly evolves successive generations of protocols. This culture, and the process for rapid evolution, are part of the regime prescribed by the new view – this, despite the US avowed commitment to the competitive ideal.

But can we take the Internet as an exemplar to understand mainstream policy in the US?[23] Consider then the FCC, the most visible of the several (contending) *mainstream* institutions which set telecommunications policy for the US.

In two major cases recently – for HDTV and then again for a spectrum plan to suit the low-earth-orbit systems (LEOS) [24] systems-of-satellites proposals – the FCC has followed *directly* the dictates articulated in the new view. For HDTV the FCC took fiercely competing rivals and, over time, encouraged and insisted that they form a single group, so that they melded the various ideas from their period of experimentation into one Grand Alliance. That is precisely the point of inflection in the new view, where as a group the industry settles on a standard (a standard with ongoing local variety, which in this case is the provision for both interlace and progressive scan). The FCC did the same thing with the LEOS contenders, to settle differences which had stymied a spectrum plan.

The US itself – the heartland of competition *ideology* – in actual *fact* is pursuing the tensioned policy that gives dynamic billing to consensus and competition together. Of course, dynamic structure as explicit policy is one result.

6.3. Asia?

So far, Asia has seemed not to take leadership in the globalisation alliances. In fact NTT has pursued a policy of being steadfastly open to each global conduit. As already noted, NTT, in partnership with others, has to date arranged to carry both Concert and WorldPartners. This adherence to peer status for the actors is the cornerstone of the prescribed power relations. On the global stage, NTT, though quietly, is providing leadership in the operation of community [25].

It is here, in Asia, that the first steps toward a true global telecommunications community of competitors may actually be under way [26].

Dynamics, in this new view, are integral to telecommunications industry structure, iterated dynamics that are specified explicitly by policy.

Notes and references

[1] To suppose that only traditional telecommunications companies will shape the global network future is, itself, a large proposition to swallow. There are several other *categories* of contenders for the global future:
- the several proposed multi-satellite networks, such as Iridium, intend to be quite global;
- with convergence, the rise of global *broadcast* networks, such as from Rupert Murdoch and Ted Turner, must be considered;
- global computer service companies ('system integrators'), such as EDS, will play a role;
- the US RBOCs, although in AT&T's shadow, will also continue to 'reach out' around the globe.

And there are other traditional telecommunications companies with an important global presence, most notably Cable & Wireless. However, the five major telecommunications companies – Deutsche Telekom, France Télécom, BT, AT&T and NTT – appear, by dint of their starting positions, to command the high ground for the next steps to a global regime. (For example, the three alliances currently carry 10, 18 and 26% respectively – or together 54% – of the world's international traffic [27]). Whether their lead *continues* will turn very much on the choices which each makes.

[2] The Deutsche Telekom-France Télécom-Sprint alliance has the working name Phoenix. However, that will not be the final operating name. This chapter uses the names of the three partners together, to identify the alliance.

[3] Concert's other distribution arrangements, to the present, are in Europe:
- Denmark: TeleDanmark
- Finland: Telecom Finland
- Norway: Norwegian Telecom

In two countries, Concert will distribute through two BT ventures:
- Germany: A joint venture with Viag
- Spain: A network purchased from Banco Santander

[4] Discussion with the president of WorldPartners Company, Simon Krieger.

[5] Unisource and AT&T have settled on a 60/40 joint venture – named Uniworld – to hold the WorldPartners European equity and operate the European entry in WorldPartners.

[6] Telecom South Africa has announced a memorandum of understanding with WorldPartners.

[7] For KDD, that is VirNet; for NTT, Members Net.

[8] Two recent papers[14] [28], provide detail.

[9] For its illustrations, this chapter turns to successive regions of the globe, and cases appropriate to the particular point.

[10] I hatched the idea myself in 1986. It also emerged elsewhere, in parallel, but I do not know the timing in the other cases.

[11] M. Shurmer & G. Lea G., Telecommunications Standardization and Intellectual Property Rights: A Fundamental Dilemma? Twenty-Second Annual Telecommunications Policy Research Conference, Solomons, Maryland, USA, 1994.

[12] Neoclassical thinking has evolved, it is important to note, various conceptual devices in its effort to cope with these phenomena. Among them are: externalities; scale, particularly in demand; monopolistic competition and economies of scope; and agency, particularly as expressed in asymmetric information

flows and moral hazard.

[13] Throughout, I unfortunately slight France Télécom's vision and role – the fault is mine, from inadequate information about France Télécom's current position in the partnership. To a lesser extent there is also a bit of information missing on BT and AT&T.

[14] D. Allen, Competing Visions for World Telecommunications: The Global Evolution of Industry Structure. In O. Grandstrand & E. Bohlin (Eds), The Race to European Eminence. Elsevier, Amsterdam, 1994.

[15] Some would not find the outcomes in Europe satisfactory, and so demur from such an example. But the whole picture must be reckoned – for one small instance, the large lead which Europe's GSM enjoys in numerous contested digital cellular markets. Also, later examples in this chapter find these sorts of arrangements decidedly at work in other parts of the globe.

[16] 'Re-industrialising' is a shorthand dubbed onto the Philippines as it (tries to) emerge from being the 'sick man' of Asia, and South Africa as its polity constructs a new era. Although KDD is the quintessentially established carrier, NTT is the dominant force in the Japanese market. Although Telia in Sweden, KPN in the Netherlands, Swiss PTT Telecom in Switzerland, and Telefonica in Spain are each the established ex-PTT (or (likely) soon-to-be ex-) in its society, Unisource is today the entrant in Europe. (AT&T *itself* is the incumbent, but divested, holder of just a long distance franchise.)

[17] Also: Concert's turn to local sales agents stands in direct contrast with WorldPartners' taking local companies into the fold. In the new view, these are different strengths of linkage, at comparable levels of nested aggregation. To do that justice should make the new ideas operational, giving them concrete purchase; it also begins to mark the differences with, for instance, Coase/Williamson concerns – all the subject for an appropriate follow-on paper. (Quality of interconnection – the force driving this chapter's logic – is even at stake. Concert maintains that its approach to structure is superior, precisely on grounds that a unified network does not suffer the glitches native to WorldPartners' 'correspondent relations', as Concert sees it.)

[18] With the Uniworld joint venture in Europe, WorldPartners has taken first steps toward *joint* service for a piece of geography. This hints at the geographical nesting discussed just below.

[19] The years-long birthing pains for a telecommunications community in Europe could also be 'spoiled'.

[20] Finland's history of strong localities – the definitive stamp of community – also parallels the new view that strong regional identities are essential for a robust global regime, for instance a strong EU.

[21] We see now why dominance is not the spoiler, in the new conception. Power relations are explicit in it, so the question of dominance is dealt with already.

[22] O. Stehman, European Telecom Policy at the Cross-roads: From Service-based Competition to the Liberalization of the Infrastructure. International Telecommunications Society, Tenth Annual Conference, Sydney, Australia, 1994.

[23] Some mainstream *theorising* about the Internet could parallel the new view. With their call for 'smart markets,' Hal Varian and Jeff MacKie-Mason [29] have broached again that the economic margin in networks exists only for the long run. Smart markets would be an auction for priority access during congestion; that could signal for capacity increases. *If* technological change is involved in the capacity increments, smart prices become part of the signal for an inflection to new consensus. Further, by *not* calling for per unit pricing, they leave scope for the 'community pricing' now largely in use, where a surrogate for community, the user's institution, pays a flat rate (whether Varian and MacKie-Mason accept either interpretation).

[24] Besides the several low-earth-orbit systems planned, there are also proposals for both mid-level and (high-level) geostationary services.

[25] In the few days since this chapter was finished, NTT has upgraded its relationship with WorldPartners to provisional member, with a trial period of six months. At the same time, NTT maintains its link to Concert, albeit the weaker sales-agent connection. Although this chapter's conclusions remain the same, NTT has stepped across a fundamental divide in the evolution of global industry structure. That impels the next iteration in this thinking.

[26] Strong regional identity for Asia is another matter again. If that occurs, it will be some years away.

[27] R. Cawley, Global Telecommunications Service: The Changing Face of the Trans-Atlantic Telephone Market. Twenty-Second Annual Telecommunications Policy Research Conference, Solomons, Maryland, USA, 1994.

[28] D. Allen, Beyond Competition: Where are We in the Dialog about Policy for Telecommunications? In D. Lamberton (Ed), Beyond Competition. Elsevier, Amsterdam, (in press).

[29] H. Varian & J. MacKie-Mason, Pricing the Internet, the World Wide Web at http://gopher.econ.lsa.umich.edu, under Economics and the Internet/Pricing, 1994.

Telecommunication:
New Dynamics and Driving Forces
A. Bianchi and G. Richeri (Eds.)
IOS Press, 1996

Chapter 4

Investments in convergence: will they return one day?

Franco Morganti
Databank Consulting, Milan, Italy

Abstract: Different approaches to convergence between telecommunications and television are considered: asymmetric digital subscriber line, hybrid fibre-coax and fibre-to-the-home, provided by a telecommunications operator. Their investment costs per subscriber are compared, as well as the potential markets and revenues for video-on-demand, pay-per-view or near-video-on-demand. A simple return on investment model, in financial accountancy terms, is then applied to the three technological approaches described, given a certain penetration Gompertz curve. The return is very poor, except in the case of near-video-on-demand services secured by analogue technology. On the contrary, a cable-television operator could be very profitable by investing in basic telephony, where the new game of competition, in convergence, will be played.

1. Foreword

Paradoxically, the whole debate on telecommunications economics can be reduced to one major question: is the telecommunications service an industry where some kind of monopoly is a natural occurrence? The answer to this question has been controversial, or even biased. This question faced both the British and the Japanese governments, in 1981, when they aimed to liberalise telecommunications in their countries to fit their liberalistic thinking [1]. It also faced Judge Greene who had to adopt the 'modified final judgement' (MFJ) in 1982 [2] [3], and the European Commission in 1987 when it decided to publish its Green Paper [4].

This chapter does not aim to resolve such a complex question. However, one issue in the debate is the introduction, and the consequent liberalisation, of new services in addition to the basic services of voice and data transmission and switching. Exactly the same issue arises with the convergence between telecommunications and television, to produce what is known as 'interactive TV' (ITV) or, from the name of the most popular service, 'video-on-demand' (VoD).

This chapter analyses this particular issue from a viewpoint of financial accountancy. Economists will recall that the reason for the dismantling of the Bell System lay in the 'bottleneck monopoly' over local access and in the fear that a dominant position could favour cross-subsidies, as in fact happened when AT&T dramatically reduced their prices on long-distance lines to compete with MCI. However, in spite of the MFJ, there remain some major outstanding problems, for example, the trend of prices in local and long-distance calls after the divestiture, the failure to demonstrate, from an econometric viewpoint, the existence of a natural monopoly in a telecommunications network and the consequent need for protection to avoid inefficiencies, and the bypass of local monopolies by geodesic networks. These suggest that an economic approach could make a helpful contribution to this controversial debate.

2. Different approaches to convergence

Consider a telecommunications operator aiming to use its network to provide video services to its customers. Three different technologies are available:

1) Asymmetric digital subscriber line (ADSL), joined to the international compression standard, MPEG (Moving Pictures Experts Group). This is a system that allows the pairs of copper wire which convey voice telephony services, to convey video. The video signal is compressed into a digital form, which allows the wire to carry more information, and the system carries a disproportionate (asymmetric) amount of information in one direction – from the operator to the subscriber's home – to provide video signals, compared with the other direction – from the subscriber back to the operator's video play-out unit – to order a programme and to fast-forward or rewind it, as with a VCR. Another signal can carry two-way voice telephony, allowing subscribers to make phone calls while using video-on-demand. Any subscriber within about five kilometres of a telephone exchange can be connected to the system. In the UK, for instance, this means about 90% of households. The ADSL system has the advantage of using existing copper-wire networks, involving subadditive cost functions. However, for two reasons, this is also the weak point: copper wires are often old and of poor quality (if they have to be replaced the investment would be higher); and when bundled together, they can interfere with video signals. The average investment cost of connecting a customer is about US$2,000, including the cost of a set-top video-on-demand decoder, the remote-control handset and the apportioned cost of ADSL system with the video-server. This average is very rough, but plausible. It does not take into account the possible cost of replacing the copper wire, which could make the cost rise to about US$3,000. Some consider the average figure of US$2,000 to be somewhat optimistic, and a more realistic projection for the year 2000. On the contrary, US estimates nowadays are lower than that average. British Telecommunications' announcement (October 1993) of an investment cost of US$600 per customer has been considered unrealistic or based on a more limited version of the system.

2) Hybrid fibre-coax, which involves building optical fibre further into the local loop and then using coaxial cable into the home, as in a cable-TV system. Here again, the investment cost per customer connected can be only roughly estimated at US$1,500, including a set-top decoder to convert digital signals for existing analog TV sets. Again, US estimates are lower: Pacific Telesis has recently announced a figure of US$800 to the FCC, without a set-top decoder.

3) Fibre-to-the-home (FTTH), for a real broadband cable connection. The estimated investment cost per subscriber connected varies from US$2,250 to US$3,000, including the set-top decoder. Technological forecasts say that fibre costs will probably drop more rapidly than the hardware and software costs involved in the ADSL solution, making FTTH competitive in the near future. This is why payback times, in the adoption of so-called 'immature technologies', have to be considered over no longer than ten years.

Of course the cost comparison between various technologies would have to be made along different penetration curves. At low rates of penetration of connected subscribers over passed homes (below 20% as the order of magnitude), ADSL becomes competitive; conversely, fibre efficiency increases with penetration rate. The figures illustrated above refer to an average penetration rate of about 30%.

All three of the technologies summarised here offer the same range of applications, with a high degree of interaction. Another approach can be adopted by telecommunications operators in countries with limited or weak regulation. This is a 'gradual', or 'step-by-step' approach, which gives customers, as a first option, a limited degree of interaction, for example, a simple cable-TV (or live TV) service, or the so-called pay-per-view TV, or the so-called near-video-on demand (NVoD), where for instance, the same film is shown starting every five minutes. The technology adopted is certainly that of the cable-TV operators with the present hybrid configuration of fibre-coax. If we assume that the operators have already installed fibre in the local loop or at least to the curb, the investment cost of providing such a service to the customer is about US$600, in case of an analog signal, or US$850, including the set-top decoder, in case of a digital signal. In the UK, where regulation is considered strong, British Telecommunications is still forbidden, by the Independent Television Commission, to provide this kind of service, because the licence for video does not permit programmes to be sent simultaneously to two or more homes, in order to protect the cable operator from the potentially dominant position of the telecommunications operator. The fact that two homes, in a VoD configuration, see the same programme at the same time, is not considered to be infringing the licence clauses because it can happen coincidently.

However, in other European countries where regulation is weak, or where the telecommunications operators are strong, which amounts to the same thing, the operators could be allowed to provide a video service of poor interactivity, even in consideration of the next telephone liberalisation of 1998/1999, where the cable operators, in principle, will be allowed to provide telephone services to their customers. This approach would allow telecommunications operators to graduate their investments, firstly by upgrading the network with a fibre-coax configuration, and secondly by investing in video-servers. Of course this kind of solution pushes operators towards the hybrid fibre-coax approach and discourages the adoption of the ADSL configuration.

3. The services and their respective markets

Many studies have been conducted recently to estimate the size of the potential markets for the various services. For instance, the services proposed by Bell Atlantic, under the trademark of 'Video Dialtone', comprise live video, pay-per-view, movies-on-demand, video games, interactive data-banks, tele-working, tele-shopping, mail order, multi-media libraries, distance education, tele-medicine, etc. As recent investigations show, video programmes are the most popular. For this is the reason, we concentrate on video, with two possible simplified scenarios:

1) A full VoD scenario, where families select an average of 140 films per year, to be ordered at any time they choose but mostly during evening and night time, for a total revenue per customer of about US$500. It should be noted, however, that up to now tests conducted in the US give consumption rates of from a quarter to a third of this figure, although a certain learning period will certainly take place and, in addition, the operators will probably apply marketing devices, such as exclusivity for some of the movies screened. This assumption would have to be supported by an in-depth market analysis, but in the circumstances it could hardly be conducted with customers as yet unfamiliar with the services on offer. The charge could either be shared between an equipment-rental charge and a cost for screening each film, or simply be applied to the film screened. In the latter case, it would mean about double the cost of hiring a video cassette for the home VCR, but with better quality, given by the broad band, broader choice and more up-to-date programmes.

2) A simple NVoD scenario, where the consumption per family is the same, but at half the price, because of the reduced interactivity of the movies programmed. A total annual revenue of about US$250 per subscriber is consistent with competitive standard cable-TV or satellite package rental (currently from US$20 to US$30 per month).

4. The return on investment model

In order to investigate the return on investment (ROI) of the various situations, a parametric exercise was applied, where:

- the potential market for video services over a period of ten years, from 1996 to 2005, was estimated by a recent study conducted in Italy [5]. The market at the end of the period was estimated at more than two million subscribers, growing from zero in the first year through a Gompertz curve, governed by steps of technology availability. The exercise can equally be applied to larger markets (BT estimates its market at eight million homes by 1999) because investments, costs and revenues are averaged, to a great extent, on the single subscriber;
- half the investments are considered as fixed: one-third in the first year of operation and two-thirds at the fifth year;
- additional gross margins per subscriber are obtained by subtracting, from the revenues mentioned above, a yearly maintenance cost of US$60 per subscriber in the case of VoD and half of this in the case of NVoD (even if the operators do not have any expertise in managing 'tree and branch' networks), and a royalty of 20% on the revenues as rights to the programme producer. The latter estimate is very rough and can be considered a minimum for today and a possible figure when the market grows. It is drawn from interviews with experts and operators. This area of uncertainty could be reduced by joint ventures between the telecommunications (now video) operators and the 'majors' producing the movies and owning the rights. However, it appears that, country by country, the commercial policy of the producers, including the management of rights, is in the hands of local distributors. To give some figures: the annual additional gross margin per subscriber is about US$340 for VoD and US$170 for NVoD.

Possible advertising revenues have not been considered in the model for two main reasons:

- VoD is also considered to be a means of avoiding annoying interruptions during screening;
- telecommunications operators would have to develop new expertise in the management of programmes with a mix of entertainment and advertising, which is the role of cable operators.

However, advertising as a source of financing VoD cannot be neglected (even if Morgan Stanley predicts only some US$0.3 per subscriber per month), especially if the results in terms of return are not unsatisfactory, as we shall see.

5. The results

Only the poorest technology available for VoD (hybrid fibre-coax) has a positive, but modest return over ten years, with an internal rate of return (IRR) of less than 1%. Both the ADSL and the broadband scenarios do not show a return over this period. The NVoD scenario, in the case of a smaller investment (analogue technique), has a positive, even encouraging, IRR of 7%, while it has no return in the digital case. This does not mean that telecommunications operators will not invest in video. On the contrary, they are consistently investing in video. On the one hand, we should not forget Chapter 12 of Keynes's 'General Theory', where very few real investments are considered to be dictated by calculations of projected returns. On the other hand, technology has often accelerated forecasts and this could be a gamble, or a way of partially financing the cost of an investment which will be made in the future anyway: that of installing fibre in preparation for broadband. But generally, in monopolistic countries, large investments with long paybacks were always tackled by cross-subsidies. Telecommunications operators are now saying that the simple replacement of the traditional copper network by a fibre-coax one can reduce telephone-network maintenance and management costs to an extent which alone justifies the investment. But in a document from Pacific Telesis, I found this explanatory sentence: 'This approach can also leverage the economies of scale representing significant entry barriers to competitors, in any of the individual segments.'

Convergence is not only a matter of telecommunications operators entering the video market. In countries where the cable market has a significant penetration, there are cable operators interested in entering the telephone market. This is already possible in liberalised countries and would appear to be possible in the whole EU in the 1998/1999 perspective. To evaluate the possible outcome, the model described above was applied (even if the application of the same Gompertz curve is largely questionable), under the following assumptions:

- that the investment cost (from literature) of providing telephony to cable subscribers would be about US$1,000 per subscriber;
- that the telephone tariff would be cut by 25% to compete with telecommunications operators, and traffic per subscriber would be in line with the average for the industry, thus giving a gross revenue of US$500 per subscriber (even higher when the club effects of the network are displayed consistently – Morgan Stanley has recently predicted some US$680);
- that the maintenance costs of the switched network will be 25% of the gross revenues, thus leaving a annual gross margin of US$375 per subscriber. Other estimates are more pessimistic, but they probably don't take into account the dramatic drop in costs arising from the fibre-coax architecture.

The exercise gives a significant IRR of about 20%. The good result is also due to the increase in the penetration rate of connected subscribers over passed homes, which is expected to increase from about 30% to about 45%.

This result is interesting, because it shows that competition could be positive without visible inefficiencies, in spite of claims of natural monopolies in telephony.

Twenty years ago, the rise of another aspect of convergence, that between computer and telecommunications (telematics) was responsible for a strong attack on the outstanding monopolistic positions. In the case of telematics, the defence of monopolists was based upon the menace of cream-skimming; now in case of video-on-demand the attack is directed to the heart of the telecommunications system – the basic telephony. Monopolies are certainly well aware of this and are trying to respond with a full range of initiatives. Twelve years after 'The Fall of the Bell System', to quote the title of a famous book, it is time to start writing the story of a new battle, if not of a new war.

Notes and references

[1] N. Curien & M. Gensollen, Economie des Télécommunications. Ed. Economica, Paris, 1992.

[2] W. J. Baumol, J. C. Panzar & R. D. Willig, Contestable Markets and the Theory of Industry Structure. Harcourt Brace Jovanovich, New York, 1982.

[3] P. W. Huber, The Geodesic Network: Report on Competition in the Telephone Industry. US Department of Justice, Washington D.C., 1987.

[4] Commission des Communautés Européennes, Vers une économie éuropéenne dynamique: Livre Vert sur le développement du marché commun des services et équipements des télécommunications. COM (87) 290 final, Brussels, 1987.

[5] Teknibank, Lo scenario e le implicazioni della multimedialità nelle telecomunicazioni italiane. Teknibank, Milano, 1993, mimeo.

Preparing for 1998

III Who Will Interconnect and with Whom?

Telecommunication:
New Dynamics and Driving Forces
A. Bianchi and G. Richeri (Eds.)
IOS Press, 1996

Chapter 5

Liberalisation, universal service and interconnection

Andrea Costa and Laura Solimene [1]
Università Cattolica del Sacro Cuore, Milan, Italy

Abstract: Due to the uncertainty over the path towards liberalisation, its extent and timing, the socio-economic impact of competition is not an obvious issue. An important barrier to competition is still the restriction on the competitive provision of infrastructure. This chapter analyses interconnection policies which will be a key element of regulation. There are two main issues: technological and economic. The definition of the interconnection price is also crucial. Different options are discussed, underlining the need for non-discriminatory conditions and the implementation of transparent cost-accounting systems. Finally, two other important issues raised by interconnection are discussed: the financing of the network infrastructure and the provision of universal service.

1. Introduction

The telecommunications sector is undergoing a tremendous change. In the last 20 years it has changed more dramatically than in the preceding 90 years since the invention of the telephone.

The interaction between technological progress and regulatory policy has been a topic on the agenda of telecommunications experts for many years. Some recent developments in economic theory provide interesting insights into the traditional approach to the natural-monopoly argument for state control. There are many reasons why these arguments no longer apply or have less force: first of all, telecommunications is no longer a commodity product with no brand differentiation. Secondly, the requests of interconnection between competing systems are increasing and they will grow as more and more networks proliferate in today's telecommunications environment and, finally, the traditional view of economies of scale does not account for the possibility of diseconomies as the telecommunications operators (TOs) grow.

Over time, the boundaries of monopoly provision have receded and the majority of member states of the European Union (EU) have taken a positive attitude to the concept of competitive provision of telecommunications services. However, an important barrier to competition is still the restriction on the competitive provision of infrastructure. Unless a single firm provides all the services, disjointed networks must interconnect to ensure consumers a complete set of services. Therefore, in the future interconnection policies will be a key element of regulation. This issue is also crucial to the possibility of providing universal services and, in this respect, the need for placing some conditions on the competing operators becomes obvious.

In the first part, this chapter briefly analyses the current regulatory situation in European Union telecommunications and focuses on the many players and activities in the liberalisation process. Then, it discusses the principal issues related to interconnection and universal service and it describes how they have been addressed in some countries. Finally, some comments are made.

2. Regulatory reform in European Union telecommunications

2.1. Players and activities in the liberalisation process

In order to clearly understand the changes in the telecommunication industry it is important to identify all the players, actual and potential, involved in the liberalisation game.

As a first step we can consider five different broad categories of players:

- customers
- regulators
- the financial community
- suppliers
- service providers

In a deregulated environment the boundaries between these categories are not always a clear cut, ie, many customers are also service providers. Large business customers will try to by-pass the telephone company and build or lease their own facilities. It is also reasonable to expect an increasing demand from residential customers for the advanced network services available on the public network.

Regulators will have a crucial role – they will design the path to be followed within the liberalisation process. In particular, they will identify the guide-lines for the separation between different types of network services, enhanced or basic, as well as the boundaries between broadcast, cable and the traditional telecommunications services. Last, but not least, a revision of the regulatory requirements for costing techniques, used to identify and allocate the total cost of providing each type of service, will be necessary to bring tariffs more closely into line with costs.

The financial community could provide the necessary financial means to meet the investments requirements coming from the liberalisation process, both directly, by participating in joint-ventures with telecommunications carriers, and indirectly, by sustaining the development of newcomers to the telecommunications industry.

The suppliers are quite a broad category, involving, among others, telecommunications and information-technology manufacturers, motion-picture production and programming, electronic mass-media as well as other niche suppliers. Traditionally, the most important supplier has been the manufacturers of large-scale telecommunications networks, including wire-line, radio and satellites.

The service-providers category will include public and private telecommunications operators, broadcast and cable television, satellite and enhanced-service providers. Moreover, there are the so-called re-sellers, firms engaged in network-service ‘packaging’ and all sorts of secondary markets for communications services, including the re-sellers of voice, data, and video services. Finally, the move from national monopoly to a more competitive environment will see the coexistence of the incumbent operator and the new service providers. On one hand, the incumbents will still dominate the market both in terms of subscribers connected and volume of traffic, and through their traditional inertia they will tend to slow down the process of change. On the other hand, the new entrants will attempt to emerge and to reach a more favourable environment. In this phase, which some European countries are actually in, the conditions of access and use of telecommunications networks are important to generate a truly competitive market.

In such a multi-player environment, not only does each of these players have objectives that may conflict with those of the others, but even within each group of players there is often a considerable diversity of attitude. Therefore, different players and different objectives will require a careful balancing. In this respect, we agree with those that have defined the 1980s as the decade of liberalisation and competition, and even if the process is still on its way, the 1990s will be the period of regulatory reforms. The regulatory conditions

Table 1. Telecommunications industry players and activity areas

PLAYERS		Infrastructure: Telecom	Infrastructure: Broadcast	Cable	Satellite	Voice	Data	Video	Telecom Services	Entertainment Services
SERVICE PROVIDERS (Telecom)	Public TOs	x		(x)	(x)	x	x	x	x	(x)
	Mobile	x			(x)	x	(x)		x	
	Private TOs	x		(x)	x	x	x	x	x	
	Enhanced serv. providers			(x)		x	x	x	x	x
	Resellers					x	x	x	x	x
SERVICE PROVIDERS (Video)	Cable TV	x		x	x	(x)	x	x	(x)	x
	Satellite TV	x		x	x			x		x
	Broadcast TV	x	x	x				x		x
SUPPLIERS	TLC manufacturers	(x)	x	x	x	x	x	x		
	IT manufacturers	(x)					x	x	x	
	Eletronic mass media			x		x	x	x	x	x
	Motion, Picturing – Programming		x	x				x		x

Note
x indicates the actual prescence of the business
(x) indicates the possibility of entering the business

Source: adapted from Egan, p 82 [2]

will be determinant in smoothing the path towards competition and for market expansion. Therefore, a crucial role awaits the regulators, especially in view of the liberalisation of infrastructure provision. This does not mean the creation of tightly-regulated utilities, because the future for TOs depends on flexibility. Accordingly, policy-makers must move with the dynamics of the sector and provide very few and clear rules to protect customers, who still need protection, while still liberalising as much as possible.

Table 1 shows that the boundaries between the groups of players are not always clear cut, especially looking at their business-area coverage. The left-hand column of the table describes the players, and the top row lists the areas in which each group is generating business.

2.2. *Implementation of European Union legislation*

The EU telecommunications policy was launched in 1987 with the publication of the Green Paper on Telecommunications and the three basic principles were:

- full opening of the telecommunications terminal-equipment and service market;
- institution of regulatory authorities formally separated from carriers;
- harmonisation of the condition of access to and use of telecommunications networks and services.

The Green Paper has been followed by many other directives and decisions associated with resolutions, recommendations and guidelines, that in recent years have smoothed the path towards a more competitive industry. There are delays in the effective implementation of the EU legislation. While some member states have introduced further liberalisation, there are others in which the incorporation of EU directives into national law has been very slow.

The actual situation is that, in general, the provision of public voice-telephony services and telecommunications-network infrastructures are still reserved for operators with exclusive rights. In many respects, all other services and equipment markets have been opened up. Not all the member states are in the same situation, but competition is emerging. The market for terminal equipment has been opened up, and to harmonise standards at community level a mutual recognition of type-approval of terminal equipment has been accepted. In the field of procurement by public utilities, where the monopsonistic and oligopsonistic nature of the supply market for equipment gives the operators an opportunity to exercise a strong influence on the suppliers, the drive towards liberalisation has been aided by the 90/531 directive in the so-called excluded sectors. The directive aims to ensure transparency, objectivity and non-discrimination in the award procedures.

Moreover, the separation between operational and regulatory functions of the PTTs is a reality in most countries, while in others the definition of the role and the responsibilities of the regulator is still an open and controversial issue.

On 28 June 1990, the Council of Ministers for Telecommunications adopted a directive on the establishment of open network provision (ONP). ONP is a regulatory framework to specially address the issue of the conditions for open and efficient access to and use of telecommunications networks and services [3]. The ONP framework directive has been followed by a directive on leased lines and by recommendations on packet-switched data services and ISDN. In the leased-lines proposal, two crucial aspects have been underlined: the usage conditions and the tariff principles for all leased lines. First of all, it established the right for users and service providers to obtain leased lines on transparent and non-discriminatory conditions for their own use or for the provision of services to third parties. In other words, no technical restrictions shall be introduced for the interconnection of leased lines and public telecommunications networks. A set of types of leased lines is defined and must be available throughout the EU with a harmonised technical specification [4]. With regard to the non-discriminatory conditions, the proposal outlines the basic principle for the implementation of transparent cost-accounting systems able to prove cost-oriented prices.

Today, tariffs for leased lines vary amongst member states, suggesting that prices do not reflect costs and, therefore, that the implementation of the article in the ONP leased-lines directive on the tariffing principle is very slow.

The ONP principles will also apply to voice telephony and cover both the service and the networks over which the service is provided with respect to the right of connection and the use of public networks. Much of the work has been done, many elements have been discussed and a proposal for a directive on the application of open access to the voice telephony is in process. Subsequent to the Commission's Telecommunications Review of 1993, the Council has agreed on the full liberalisation of voice telephony by January 1998 (Resolution 93/C213/01). There will be transition periods up to five years for Greece, Spain, Portugal and Ireland, while the transition period for Luxembourg will be two years.

The other key issue is competition in the provision of infrastructure. A Green Paper on the telecommunications-network infrastructure and cable-TV networks and their future liberalisation/regulation will soon be published [5]. The challenge for both national and European policy-makers is to craft a regime that strikes a balance between the desire to ensure efficient infrastructure investment and to accord the TOs sufficient pricing flexibility to respond to competition as it emerges.

The main concern is that the liberalisation of infrastructure provision will affect the controversial problem of cross-subsidy arrangements actually embodied in the tariffs, in order to meet the obligation to provide services at an affordable price to all members of the community. This gives the regulators a further dilemma: how to combine the benefits of competition in areas where multiple competitors cannot be supported with the constraints needed to avoid, in such areas, the exploitation of customers.

The universal-service concept underlines three main issues:

- the minimum service (at a given quality)
- provision to all users
- affordable pricing

Universal-service obligations generally require operators to provide a certain basic service to customers whom they may not have sufficient incentive to serve. The 1993 Council resolution (93/C213/01) identified as one of the main objectives of the EU telecommunications policy, the liberalisation of all voice telephony 'whilst maintaining universal service'. The latter is a general statement. It is, therefore, crucial to identify the main elements of universal service and the principles with regard to its financing. In a competitive environment the revenues once used to support universal service may no longer be available. Moreover, its maintenance and development in the new competitive environment must go hand-in-hand with the process of tariff adjustment towards underlying costs. Furthermore, changes in technology and demand give less strength to the concept of universal service as a benchmark for policy decision.

The EU fully supports the need to maintain and expand universal service and the obligations will be developed in line with the basic principles of ONP. Within the EU, all member states already place a variety of obligations on TOs and service providers. Given these, sometimes, divergent measures, it becomes important to identify common principles to ensure a comparable level of universal service throughout the EU [6], without jeopardising the freedom of each state to impose complementary obligations, when a particular national situation calls for it.

Achieving consensus on the issues we have just examined will still require substantial discussion among the players. However, there is increasing evidence that competition in the provision of infrastructure does not need to undermine the possibility of providing universal service. This issue will be discussed in the next section.

3. Interconnection and universal service

In several regulated industries the incumbent firm must supply an 'input' (local access, transportation of energy, airport facilities, ...) to a competitor in order for the latter to compete on the final output market. Access is a very particular 'input' and it is difficult to identify a theoretical framework in which to include all its facets.

The main concern of the regulatory authorities is fairness of access and there have been two main responses to this concern: the imposition of mandatory interconnection, and the break up of the incumbent firm in order to make impossible, or simply more difficult, unfair treatment of the rival. These two responses, even if not exactly, identify the positions adopted by European (the first) and American (the second) regulators. We will leave aside the second response and concentrate on the first one, especially on the determination of a price, for an input that is so difficult to define.

The interconnection question is related to 'access' to the final customer. Broadly speaking, there are two main interconnection issues: one technological and the other economic.

Technical issues refer to standards as well as levels and types of interconnection.

Standards have been set up in the telecommunications world since the beginning of this century. In fact, to handle an international telephone call requires, by definition, a common standard for a successful connection. Considering the actual technological environment, the question is not simply related to a flow-through of voice message, as in the past, but also to data, with different technological environments, from analogue to digital, and with many options, optical cables to microwave transmission to cellular and so on, that ultimately will affect the transmission quality.

Standardisation will be a leading issue in the European Union policy [7]. In fact, by adopting the ONP approach, standards have to be established in order to harmonise and define the requirements for interconnection.

The level and type of interconnection are not only related to technical issues but they directly affect the level of competition, actual and potential, coming from new competitor(s) to the incumbent. Connections can be made at local, trunk and transit exchanges and the level of connection can influence the new entrants behaviour because:

a) It defines the infrastructure needed by new competitor(s). The highest (lowest) level of connection would be required to the incumbent network, the lowest (the highest) infrastructure requirement would be indirectly imposed on new competitor(s). If interconnection could be granted at the transit exchanges (the highest level), a few exchanges, built up at crucial points, would be enough for new competitor(s) to handle all traffic to every customer in a country or in a defined area [8]. On the other hand, by requiring connection at the lowest level, the creation of a comparable network (local, trunk and transit) would be imposed on new competitor(s). Depending on the infrastructure requirements, a different kind of competition could emerge.

b) It defines the transmission quality. For instance, the conversion digital-to-analogue can drastically influence the error rate as well as slow down the transmission speed, so reducing transmission quality. Because part of the local exchanges is still electro-mechanical, connecting these switches will, by definition, impose a digital-to-analogue conversion. However, this is a temporary effect that will disappear with the progressive substitution of electro-mechanical switches by digital ones.

A technically-related question is equal access. It can be defined as one in which customers can keep their existing local line and choose among two or more long-distance operators by dialling different but equally burdensome access digits [9]. This issue directly affects competition between carriers. In fact, a carrier with a shorter or easier numbering than a competitor, will enjoy a clear non-pricing advantage. This has been a central question in all the countries facing the liberalisation of telecommunications, from the United States, to the United Kingdom, Japan and New Zealand [10] [11].

Considering the economic issues of interconnection, on theoretical grounds there are differing opinions. Generally speaking, these opinions are built on different approaches to the nature of the telecommunications network. The final goal for each theory is economic efficiency, but diverging approaches sometimes imply very different solutions.

Some authors suggest that the problem of pricing interconnection should relate to the concept of 'appropriability', and be studied with reference to the framework of intellectual property [12] [13]. We think that this framework can be used to identify two possible opposite views on the nature of the telecommunications network, both related to the economic characteristics of access to information.

First of all, we can refer to the concept of appropriability as the degree of control over the telecommunications infrastructure and, consequently, to the need for contributing to its financing. The idea of a 'non-proprietary' network is directly connected to access at a zero price. 'Free interconnection is analogous to the case of a public industry-wide product standard while interconnection fees are analogous to licensing fees paid to achieve compatibility with proprietary technology.' (Brock, p 4) [13] In particular, Brock suggests that free interconnection will force prices to marginal costs and eliminate all kinds of discrimination. This will happen because new competitor(s) will select their competitive game by focusing on 'any particular segment of the market that is priced above cost without attempting to supply the entire network that is necessary for consumer utility' (Brock, p 23) [13]. Under certain conditions this approach will lead to economic efficiency. But this view, suggesting the idea of a 'non-proprietary' network, does not face the question of how the telecommunications infrastructure should be financed.

This issue is particularly pointed out by the second approach. The only worry of a competing service provider who purchases access to the established network, and then can resell it, is about the cost of interconnection and not about the cost of constructing the network. 'The problem of appropriability makes it necessary and legitimate for the owners of the information resource to discriminate in the price charged for access depending on whether or not the user intends to resell it. This is true even though there is no difference in the incremental cost of supplying access to a reseller or an end-user. That is, both markets are characterised by indirect appropriability.' (Mueller, p 3) [12]

The proponents of this approach, by using public-finance theory, develop taxes on interconnection, in order to subsidise universal service. This is a question generally stressed by the regulators in different countries and also by the European Commission [14].

The basic idea is that the traditional state monopoly led to a structural deficit on access, subsidised by revenues emerging from traffic, because of a general obligation to provide universal service imposed on carriers. Digital innovation increased this problem, and competition stressed the issue of how to finance the so-called 'access deficit' by limiting the possibility of subsidisation and by introducing interconnection.

In this stream, most economists approach the problem in terms of the incremental costs of supplying an individual access link to the network. More recent and sophisticated theories say that interconnection charges should also include opportunity costs, in order to recapture the contribution to network overheads and subsidies foregone by providing interconnection to a competitor. In particular, 'if a component of a product is offered by a single supplier who also competes with others in offering the remaining product component, the single-supplier component's price should cover its incremental cost plus the opportunity cost incurred when a rival supplies the final product' (Baumol, p 21) [15]. In such a way, the problem of financing a telecommunications network characterised by economies of scale and scope in the production of a variety of services could be faced.

Once agreed on the imposition of a fee, we need to identify the cost of interconnection and, in particular, it seems there is no easy solution to the problem of how to allocate the common costs of production that normally cannot be specifically attributed to any service. They generally represent the greatest part of the total costs of a telecommunications carrier. Moreover, some suggest that interconnection should be priced below cost in the early stages of competitive entry to give the new operators a chance to survive, but as the latter gain market shares the rates should be adjusted to reflect true incremental costs [10] [16]. The basic idea is that the long-term positive effects induced by competition will be higher than the costs paid by the incumbent, represented by the difference between the cost and the real (lower) price paid by the new competitor(s). In other words, they suggest that a 'subsidised/assisted' entry would allow the consumers to be better off in the long run. In particular, Vickers and Yarrow [16] suggest that the regulator has to price interconnection charges lower than costs faced by the incumbent in the local network (local loop and switching) in order to provide a 'competitive factor'.

The above description of what we have identified as the main approaches to interconnection points to the uncertainty surrounding the issue and certainly the debate is growing, especially with respect to the pricing policies, as we will see in the next section.

3.1. *Interconnection pricing*

If we assume a regulator's pro-competitive attitude, then the question is how to establish procedures to define the interconnection price. Each of the approaches described above identifies a particular interconnection pricing policy.

As we have seen there are two main alternatives: free interconnection or an interconnection fee.

Free interconnection, in some conditions, will force prices down to marginal cost, as Brock [11] has argued, because competitors will select the level of competition by focusing only on customers who are charged above cost. This solution is not only theoretical, but it has been used in the US terminal-equipment market, where no fee was imposed on terminal equipment interconnected with the telecommunications network.

When a fee is imposed we can envisage the following candidate principles for identifying interconnection prices.

First, if the incumbent can freely establish the interconnection price and it can influence its competitor's(s') profit margin by fixing high interconnection prices it will be able to exclude or minimise the competitive threat. In addition, the production level will not be optimal because the incumbent firm will produce at the level where marginal revenue is equal to total average cost (Vickers and Yarrow, pp 73–76) [16].

Second, the interconnection price can contain all the elements connected to the subsidisation of the local network. As pointed out before, this situation is related to the access deficit contribution and to its determination.

How can the access deficit be determined? On one hand, we can easily identify revenues coming from connection and rental charges. On the other hand, on costs, some problems emerge with respect the identification of the 'true' costs of access because of the huge weight of common costs on total expenses.

In accounting practice, a generally-accepted procedure for identifying costs, which is not only applied to public-utility industries, would be:

a) allocation to a specific service of all costs directly attributable on a non-ambiguous basis;

b) allocation of the remaining costs, the 'common costs', on the basis of a relevant 'allocator'. This allocator could be based on the 'production volume' of a specific service (for instance, the number of calls or minutes of transmission) [17] or on 'direct attributable costs'

[18] or on 'gross revenues'.

In the American telecommunications industry, some regulators have developed detailed accounting procedures based on the rules described above. In the United Kingdom, BT interconnection charges are based on the mechanism of fully-distributed costs (FDC), which includes relevant overheads and a reasonable rate of return on attributable assets. Because of constraints on universal service, and the related deficit in access business, Oftel stated that 'interconnection charges need to include an element for the recovery of the deficit when appropriate' (Oftel, p 4) [19].

This mechanism has been widely criticised because it does not lead to marginal-cost pricing. In practice, it leads to an average cost calculated on historical costs. Thus, pricing policies based upon it are not efficient.

In particular, the implementation of a fully-allocated costs-procedure stresses the problem of cost-allocation as well as that of price-discrimination between regulated and non-regulated markets. The latter has been a well-known issue in the American experience. The FCC itself acknowledges this trade-off when it states that 'a cost allocation system can represent a strong deterrent to anti-competitive activity and, at the same time, be so detailed and rigid that it imposes on a carrier a complex and inflexible rate structure, one that may have little relation to consumer demand. If such a rate structure is deployed in a competitive environment, it may result in distorted consumption decisions, distorted production decisions, and distortions of the competitive process.' (FCC, p 18) [20] Both imply a tendency to foster cross-subsidisation, flowing from regulated through unregulated services, and inefficient pricing.

Accounting separation between different businesses could help to identify possible cross-subsidisation. But is it enough to ensure consumers' long-term interest? From a regulator's point of view, probably not, because the carrier itself is the source of nearly all information about its costs. There is an asymmetry of information which clearly favours the regulated firm. Thus, it is not surprising that in the American case the FCC faced an historical impossibility in determining rate changes as demonstrated in the Telpak tariff story [21]. A possible solution could be the creation of different companies for each service. But at the same time, joint-production of different services affected by economies of scale and scope, particularly in the new technological environments, could help to introduce new generations of services with the best production technology.

New procedures have been developed based on the concept of incremental cost and, eventually, of opportunity cost.

Incremental costs '... arise if the provision of the relevant services occurs in the context of the production of other related services and there is the possible advantage of economies of scale and scope.' (Neu and Neumann, p 7) [10] From an economic point of view an accounting practice based on incremental costs is more desirable than a fully-allocated costs practice. However, they can be very difficult to implement [22]. There are also stand-alone costs arising from the '... provision of a service or a group of services all by itself, that is to say without the benefit of any economies of scale or scope afforded through the provision together with other services.' (Neu and Neumann, p 7) [10] These two concepts represent the lower (incremental cost) and upper level (stand-alone cost) for pricing decisions.

As third possible option, interconnection prices can be set at the same level as those of any other customer using the incumbent's telecommunications network. Thus, the potential competitor will pay, and be considered, as any other customer with a comparable traffic that uses the network. In this situation, prices will not necessary be forced to marginal cost. In fact, pricing will be related to the competitive pressure coming from other carriers as well as to the regulatory environment. Moreover, the (new) competitor will face a possible disadvantage because it has to pay customers' prices and sustain costs to provide the service to its customers.

This approach, in our opinion, is quite interesting because of the underlying concept of interconnection as a service *tout court*. But it is clearly related to the level of prices existing in the telecommunications market. Because of the present pricing situation thought out and built up in a monopolistic environment, a strict application of this methodology could provide too many advantages for the incumbent. However, in a competitive environment it could be quite desirable because it would permit the creation of many pricing structures in parallel with prices for customers. This could also lead, under some conditions, to 'maintain contribution levels necessary to keep marginal customers on the network', but it could also have a strong 'incentive to have usage prices even for services with costs that do not vary with usage in order to charge a large fee to a competitive network' (Brock, p 20) [12] and, thus, reduce the competitive threat coming from the rivals' network.

The last option recalls the definition of interconnection prices as it is in the present system for international networks. In this way, interconnection prices would be settled between two carriers at the same level. Without the regulator's intervention, the incumbent would try to design a set of interconnection prices which minimises the possible competition. Thus, prices would be set at the highest possible level in order to marginalise the competitor's threat. While, if prices were to be set close to marginal cost or at a low enough level, then the competitor would be in an extremely favourable position.

The regulator's intervention would lead, almost by definition, at least to the establishment of a standard framework of rules or, as in the American and British experience, standard charges.

Until now, the existence of different options for the costing procedure has been underlined. But for all the possible reference costs (average, fully-distributed, incremental, opportunity, and so on) a further decision has to be taken: it must be decided whether the evaluation should be at historical costs or current costs.

The answer to this question is quite important because it can lead to a higher or lower contribution to the incumbent's telecommunications network by the competitor. Historical costs will grant, *ceteris paribus*, a higher level of costs because of the old regulatory environment and the higher costs related to the electro-mechanical technology. While substitution costs will lead to a substantial reduction in costs, particularly when we do not consider marginal replacement (a digital switch for an electro-mechanical one), but the entire network.

3.2. Financing the network infrastructure and universal service

Finally, there are two more crucial issues raised by interconnection. The first refers to the financing of the network infrastructure and the second to the provision of universal service.

In general, in a monopolistic environment the network growth has been and, sometimes, still is financed by the use of rate of return (RoR) on invested capital [23]. In a liberalised market, the modernisation of the network infrastructure will be challenged mainly by carriers' shareholders and not, as in the traditional RoR regulation, by rate-payers. The increasing competition in the telecommunications industry spurred the regulators to find new mechanisms, the so-called incentive regulation, and it has been mainly adopted, at least in pricing regulation, in the form of price-cap [24].

The adoption of price regulation generates tariffs more related to costs, but it raises the problem of how to implement the provision of universal service in a competitive framework. The introduction of new services spurred by competition can benefit users, such as disabled and elderly people, as well as rural areas, but at the same time they can involve high costs and only a thin slice of the market. If we consider the social goals related to universal service as part of the telecommunications policy, then the real question is whether they can be achieved only through competition or whether it will be necessary to impose some obligations on carriers.

What is the minimum service? No doubt in the past it was, and in some ways it still is, the telephone service, but in the future, perhaps, it will be the so-called US idea of 'video-dialtone'. In any case, the final goal will be the increase in customers. Telephone penetration has not been eroded in any country that has introduced infrastructure competition. On the contrary, access to the telephone has steadily improved. In the future competitive environment, we will find not only different carriers but also different competing technologies. Then indicators for universal service will be provided by the sum of different new connections.

Even if competition seems to achieve universal-service goals, it does not seem reasonable that, especially in the short term, public authorities should not impose any obligations on carriers' licences, most of all on pricing freedom and flexibility. First of all, the possible cross-subsidisation between competitive and non-competitive markets is likely to induce non-efficient behaviours. In particular, price-discrimination might be found. In this sense the increase in the off-peak long-distance rates, both in the British and American experience, is quite interesting when there is no reason for it if carriers are pricing at the marginal cost. Second, it probably leads to a proliferation of networks, both private and public, with a lot of problems of interoperability. ONP represents a technical answer to this problem.

Finally, a general co-ordination between private and public networks will be required.

Depending on the set of possible solutions, the impact of liberalisation on telecommunications carriers will be different. The interconnection solution will affect the real level of competition by defining the available margin for competitors. In the UK, for example, the potential Mercury margin varies from zero to around 50% of its final tariff, depending on the type of connection and the time of day [25] [26] [27]. This close link has been broken by introducing more licensees and relaxing constraints on radio and cable companies to provide telecommunications services and links. (For a short discussion on interconnection practices in some selected countries, see the appendix.)

4. Conclusions

The innovation breakthrough and the regulatory behaviour have questioned the traditional approach to providing telecommunications services. Moreover, the liberalisation process is taking the future of telecommunications towards what we can call a 'multi-network interconnected environment'.

This new framework poses some questions.

The first is related to the right of interconnection: in other words, when and how a (new) competitor should be allowed to be connected to the incumbent's network. Strictly speaking, this is linked to the legal framework existing in each country and it is the preliminary condition for building a competitive environment. From an economic point of view, this issue is related to the existence of monopolistic conditions, to the regulatory behaviour and to the pro-competitive attitude of the European Union.

The definition of the interconnection price is the second important issue. The EU has stressed the need for non-discriminatory conditions and the implementation of transparent cost-accounting systems in order to bring tariffs close to costs. As we have seen, there are many cost-accounting options claiming to reflect the 'true' cost of interconnection. In any case, we believe that it will be difficult to identify such a 'true' interconnection cost and, therefore, the price that will reflect it. With regard to this point, it is our opinion that the system will end up with a 'market-interconnection price' as defined in agreements between telecommunications operators. This interconnection price will be lower when the level of competition, number of carriers and alternative systems and so on, are higher because of the competitive pressure. This could imply, at the beginning of a competitive market, the possible intervention of the regulator in order to counterbalance the incumbent's market power and define a clear framework for all competitors. But this intervention could lead to non-flexible pricing schemes and slow down the possible introduction of technological alternatives and the process of reducing interconnection prices.

At the beginning of section 3 we considered, as many do, interconnection as an input of production, even if stressing that it is a very peculiar input that sometimes underlies the concept of a 'proprietary' or 'non-proprietary' network. Initially, this approach could be the right way to consider interconnection, especially when attempting to identify all its facets and the principles governing the pricing procedures. In this phase, it seems to us that the role of regulators is crucial and, with the issues related to interconnection and cost separation, they will also have to concentrate on items such as number portability and the provision of universal service. In the long term, once the players in the new and more liberalised environment have agreed on the issue just discussed, it is our opinion that competition will flourish without special regulatory intervention and interconnection, because the increase in capacity generated by the existence of multiple networks (stressing the concept of 'proprietary' network with respect to that of 'non-proprietary' network) will become more similar to a service *tout court*, and be priced accordingly.

In our view, this supports the idea that the 1990s will be the decade of regulatory reforms. Competitors still feel that the incumbent operators have too much scope for delaying the introduction of competition, thereby fostering competition; while assuring non-discriminatory conditions is the regulators' main job. The challenge for national and European policy-makers is to craft a regulatory regime that strikes a balance between the desire to ensure reasonable rates and promote efficient infrastructure investments and the need to grant the TOs sufficient pricing flexibility to respond to competition as it emerges.

Appendix

A.1. Interconnection practices in selected countries

In this appendix we discuss some issues related to interconnection in some of the principal countries. In general, the reference is to G7 countries, Canada excluded. However, other interesting experiences will also be included.

The speed of the liberalisation process is quite different from country to country. In the last decade in the United States, Japan and the United Kingdom, a wide liberalisation process has been developed. As a result, competition expanded in different services, from cellular to telephony, which also involved a competitive provision of the infrastructure. The final result is the presence of many operators potentially carrying new and old telecommunications services. In the United Kingdom, for example, there are four GSM operators and a wide choice of cable providers as well as four telecommunications carriers. In continental Europe (France, Italy and Germany) the liberalisation path has been much slower. In these countries, in particular, the reforms have generally been carried out in order to preserve the fixed telecommunications infrastructure against possible competitors. A more open view has generally been taken for cellular, analogue and digital, where two (France and Italy) or more operators (Germany) have been licensed.

Thus, is not surprising that the interconnection issue has a different priority in the various environments. It plays a central role in the most competitive environments, while it cannot be considered so crucial in countries where competition has only recently been introduced. In US, the FCC has established an access-charge plan in order to identify the conditions for access to the local loop for all competitors. On the contrary, in the United

Kingdom, interconnection has been considered a commercial matter between parties. Oftel's intervention is permitted only if the parties fail to reach an agreement [28]. The same approach has been followed in Japan. However, an access-charge system is also under consideration in the UK. While in France and Germany, interconnection is considered to be strictly a commercial matter.

The level of competition heavily influences how interconnection takes place. Since 1991, Oftel has permitted various providers, from cable to utilities networks, to offer telecommunications services. Also, in USA, starting from 1992 the FCC obliged local companies (LECS), to supply physical space in their own facilities to competitive providers [29]. In other countries, the right of interconnection is clearly related to the telecommunications-network infrastructure run by the previous monopolist operators. In these countries, new competitors are considered, in some ways, subordinate to the main telecommunications infrastructure. This is the situation in France and Germany. In Italy, Telecom Italia [30] still retains a monopoly in most of the telecommunications services.

Generally, where the competitive environment is quite advanced, competitors can easily have access and build up their own network (USA and UK). While in other countries (Italy, France, Germany and Japan), in order to avoid a by-pass of the main telecommunications facility, there is still strict regulation. In particular, a set of very detailed rules has been imposed on mobile-carrier competitors to connect their traffic to the closest fixed exchange. These rules were also applied in the United Kingdom until the duopoly revision in 1990.

In a competitive environment, equal access is a crucial question. It has been quite a controversial issue in the American experience of the competition between long-distance companies. In particular, it was recognised that a non-equal-access competitor would enjoy 45% off the equal-access rate. This situation has spurred the local companies to innovate their facilities: the lines in equal-access offices were almost zero in 1983, 92% in 1988 and 99% now. A similar approach is now pursued in the UK. In a recent document [31], Oftel considered some issues related to equal-access competition, such as numbering, number-portability and discrimination on quality of service. In other countries, since non-discriminatory interconnection is normally provided, equal access has not been considered in such a detailed way.

Another key issue is how to relate access-prices to costs. In the UK and the USA, prices are based on fully-distributed (or allocated) costs practices, while in other countries prices have been linked, in some ways, to the final tariff (adjusted or not). It is quite interesting to note that the introduction of competition is going hand-in-hand with deep accounting analysis by the regulator. But the most innovative experiences have been developed in Australia by Austel, where directly-attributable incremental costs have been introduced [32], and in New Zealand, with the adoption of Baumol's efficient component-pricing rule. To our knowledge, these are the only experiences where historical costs are not the guideline for interconnection pricing. However, in both the UK and the USA other costing procedures are under consideration. In the UK in particular, Oftel has recently considered the possibility of basing interconnection charges on long-term incremental costs. 'However, there are some important practical and policy issues associated with using incremental costs which will require some detailed and careful analysis.' (Oftel, p 22) [31]

In the different countries, interconnection pricing practices are mainly related to the following three issues.

First, the access capacity. In this case, the incumbent normally charges direct costs for the connection (UK, USA and France). In the UK, BT is also entitled to receive 50% of total costs incurred in increasing its capacity in order to satisfy the competitor's demand. While, in other countries no formal provision exists.

Second, because of the traditional structure of tariffs, based on low monthly charges and subsidised local tariffs, monopolists faced, and are still facing, a 'structural' deficit in the access business [33]. How can this deficit be recovered and how can the new competitors take part in its financing? In the UK, for instance, BT interconnection charges are based on the mechanism of fully-allocated costs, which includes relevant overheads and a reasonable rate of return on attributable assets. Because of the constraints on universal service, and the related deficit in access business, Oftel stated that 'interconnection charges therefore need to include an element for the recovery of the deficit when appropriate.' (Oftel, p 4) [19] and introduced, with the duopoly review, an access-deficit contribution, in order to recover the pro-quota deficit in the access business faced by BT. A different approach has been followed in the USA in order to recover the so-called non-traffic-sensitive costs. These costs have traditionally been attributed to long-distance carriers, while now they are shared between users, by the end-user common-line charge ($3.49 per month), which covers more than 60% of access costs, and long-distance carriers ($0.04 per minute) for originating and terminating access. A universal-service fund will fully guarantee local companies with the highest access costs. A similar path has been recently followed by Japan. No formal provision exists in other countries. In Italy, the cross-subsidisation fund (Cassa Conguaglio per il Servizio Telefonico) created in 1981 in order to ensure local companies the financial means for network investments, disappeared as a result of the merger of the different telecommunications companies.

Finally, interconnection pricing practices are related to traffic. In this case, different solutions will apply. There exist discount tariffs based upon volume (France) and the conditions normally vary with respect to the contracting capacity of the parties and the available margin in the final market. Thus, *ceteris paribus*, cellular operators, who form a 'rich' market, will be able to sustain higher interconnection charges than those operating in fixed telecommunications services. In the UK, interconnection prices are related to time of day and distance. Now Oftel is considering the introduction of a system of access charges. An access-charge system, based upon a fixed charge per minute, exists in the USA and has also been implemented recently in Japan.

Notes and references

[1] Although this chapter is the result of common work, sections 1 and 2 and the appendix were written by Andrea Costa, and sections 3 and 4 were written by Laura Solimene.

[2] B. L. Egan, Information Superhighways: The Economics of Advanced Public Communication Networks. Artech House, Boston, 1991.

[3] The EU's ONP programme can be compared with the USA's Open Network Architecture (ONA) and Japan's Open Network Doctrine (OND). These are all equivalent initiatives; although they differ in approach and detail they are all designed to allow enhanced-service providers fair and equal access to the telecommunications network.

[4] The framework directive provides that the EU should request that the European Telecommunications Standards Institute (ETSI) draw up European standards for technical interfaces and service features for open access.

[5] A preliminary draft of the Green Paper was prepared and circulated in October 1994 [34].

[6] See the proposal for a Council resolution on 'Universal service principles in the telecommunication sector' [14].

[7] EU policy is clear on this subject. The most recent example is the Green Book on Mobile Phones.

[8] International telecommunications are a clear example of this situation. For instance, the former Italian international carrier, Italcable, had a similar network design with only a few exchanges to handle traffic and to interface with the international carriers.

[9] Oftel, Annual Report 1991. Oftel, London, 1992.

[10] W. Neu & K. H. Neumann, Interconnection Agreements in Telecommunications. Wissenschaftliches Institut für Kommunikationsdienste, discussion paper No 106, Bad Honnef, 1993.

[11] D. Walker & J. Solomon, The interconnection imperative, *Telecommunications Policy*, May/June, 1993.

[12] M. Mueller, Comments on Interconnection for European Telecommunications. 21st Annual Telecommunications Policy Research Conference, Solomons, Maryland, USA, 1993.

[13] G. W. Brock, Interconnection Conditions, Access Charges, and Universal Service. 21st Annual Telecommunications Policy Research Conference, Solomons, Maryland, USA, 1993.

[14] Commission of the European Communities, Developing Universal Service for Telecommunications in a Competitive Environment. Communication from the Commission to the Council, the European Parliament and the Economic and Social Committee, COM (93) 543 Final, Brussels, 1993.

[15] W. J. Baumol, Modified Regulation of Telecommunications and the Public Interest Standard. London, 1991.

[16] J. Vickers & G. K Yarrow, Privatization: An Economic Analysis. The MIT Press, Cambridge, Mass., 1988.

[17] This methodology in based on relative production: common costs are allocated on the basis of the number of units produced of a specific service with respect to the total units produced of all services. An example of this approach has been proposed, among others, by Mercury and Arthur Andersen during the discussion on Oftel's consultative document [19]. In this proposal, the 'allocator' was based principally upon busy-hour earnings, on the argument that the principal cost driver for conveyance is the installation of capacity [26].

[18] Common costs allocated to a specific service will be a percentage of common costs equal to their quota on direct attributable costs. The rationale of this methodology is based on the assumption that the higher direct costs are attributable to a specific service, such as materials and labour costs, the higher will be the plant use for its production and other common costs.

[19] Oftel, Interconnection and accounting separation. Oftel, London, 1993.

[20] Federal Communications Commission, Policy and Rules Concerning Rates for Dominant Carriers. Report and Order, CC Docket n.87–313, Federal Communications Commission, Washington DC, 1989.

[21] The Telpak tariff was introduced by AT&T in 1959 as soon as the FCC decided to liberalise the introduction of microwave facilities for users with 60 or more private lines. AT&T, in order to discourage private microwave adopters, filed a new tariff for private lines (the Telpak). It contained such huge discounts, in comparison with the old one, that it was immediately suspended for 90 days. In addition, the FCC started an investigation into the new tariff. In 1964, the FCC considered part of this tariff discriminatory (Telpak A and B for medium users) and, thus, cancelled it. It reached no decision about Telpak C and D for large users because of a lack of data. Finally, in 1977, after 20 years, the FCC ruled out the tariff because of price discrimination. Then, Telpak users appealed to courts in order to maintain their rights over the old tariff. The courts judged in favour of the customers (Brock, pp 206–210) [35].

[22] In general, interconnection costs have to consider:
a) costs sustained by the incumbent to provide interconnection between its network and the infrastructure of the competitor. In practice, these costs are related to the provision of physical connections between carriers' central switches;
b) the increased capacity of the incumbent's network related to traffic carried by the competitor which requires interconnection. The increased capacity results, *ceteris paribus*, from the difference between the traffic carried and interconnected by the competitor and the incumbent's decreased traffic related to the competition.
c) the costs related to handling traffic coming from the competitor.
Costs a) and b) are considered as fixed costs, while c) are variable costs. In general, incremental costs can only be calculated with respect to the first two categories.

[23] The traditional regulatory policy for the telecommunications industry, as well as for other public-utility industries, has been rate of return (RoR) regulation. This mechanism is based on the concept of cost plus, and it has been widely used in the United States to constrain the market power exercised by public utilities. The effects of RoR regulation have been studied and the principal problem of the system is the lack of incentive for cost reduction. Under this regulatory mechanism a carrier can increase its total earnings simply by increasing its aggregate investment. This can be done by expanding both the rate base and the permitted expenses. By making investments related to regulated business, or to every activity in the European experience, a firm can increase its profits without any change in its permitted RoR. Therefore, neither efficient pricing nor an efficient production seem to be directly addressed by RoR in its traditional form.

[24] The experiences of incentive regulation will be fully developed within the liberalisation process. The final result will be a reduced telecommunications bill in real terms. This reduction will be originated by a tariff rebalancing, increasing the cost of access to the network (monthly subscriber charges and local rates) and decreasing long-distance (domestic and international) as well as private-line costs. This principle has been clearly expressed by European Commission in designing the ONP's framework and

has been restated by the Commission in a recent document [14]. The liberalisation process will stress this phenomenon by emphasising price competition in services mainly devoted to business customers affected by high-tech requirements and a large use of telecommunications. This pressure is growing, first of all, on intra-European as well as international rates because of the comparisons that are possible between countries. But the degree and the speed of this trend will greatly depend upon the level of competition in the different markets.

[25] M. E. Beesley, Privatization, Regulation and Deregulation. IEA-Routledge, London, 1992.

[26] M. Cave, Interconnection Issues in UK Telecommunications. Paper presented at the conference 'Access Pricing in Network Utilities. Theory and Practice', London Business School, 1993.

[27] A. Costa, Concorrenza, interconnessione e regolamentazione: il caso British Telecom-Mercury. In Pontarollo E. (Ed.), Apertura dei mercati e regolamentazione. Il caso delle telecomunicazioni. Seat, Roma, 1993.

[28] The intervention of an arbitrator may have different forms, for example, in New Zealand the incumbent can be taken to court for anti-competitive behaviour.

[29] This rule applies to large LECS. The final goal is to limit the monopoly powers exercised by local companies in the local loop.

[30] Recently created by the merger of SIP, Italcable, IRITEL (ex-ASST), Sirmi and Telespazio.

[31] Oftel, Interconnection and accounting separation: The Next Steps. Oftel, London, 1994.

[32] Augmented by a small contribution to the incumbent's local loop.

[33] This situation has been increased by the introduction of digital technology, with a costs pattern different from that of electro-mechanical technology, related more to access than usage.

[34] Commission des Communautés Européennes, Vers un environment de communications personelles: Livre Vert sur un approche commune dans le domaine des communications mobiles et personnelles. Presenté par la Commission, COM (94) Final, Brussels, 1994.

[35] G. W. Brock, The Telecommunications Industry: the dynamics of market structure. Harvard University Press, Cambridge, Mass., 1981.

Telecommunication:
New Dynamics and Driving Forces
A. Bianchi and G. Richeri (Eds.)
IOS Press, 1996

Chapter 6

Interconnection: the key to further liberalisation in Europe

Bridger Mitchell
Charles River Associates, Palo Alto, USA
Werner Neu and Karl-Heinz Neumann
Wissenschaftliches Institut für Kommunikationsdienste, Bad Honnef, Germany
and
Ingo Vogelsang
Boston University, Boston, USA

Abstract: Interconnection has turned out to be the key to effective competition in telecommunications markets. Consequently, in a study for the European Commission, the whole variety of aspects of interconnection including technical, economic and regulatory issues were covered. This chapter presents two main portions of this study. First, we present a systematic analysis of existing experience of interconnection in Europe and other parts of the world. From this we derive some lessons regarding the proper regulatory treatment of interconnection. Second, we describe and analyse the current European interconnection policy and discuss its implications, in particular with respect to telephony in the fixed-link and mobile networks.

1. Introduction

1.1. Interconnection and competition

From some discussions on interconnection with regard to competing network operators and service providers one might get the impression that interconnection is a relatively new issue in telecommunications. That is really not true. The history of interconnection is as old as the history of telecommunications itself. From the old days of telecommunications, the predominant pattern of interconnection has been the interconnection of networks from operators that were active in locally, regionally or nationally-separated markets. The main purpose of interconnection was to enable users in one region to reach users in another region. The new dimension of interconnection in the modern history of telecommunications results from the intention of operators to be interconnected to another operator's network in the same region or market.

Many people find it difficult to understand how competition in telecommunications can be introduced and can work when initially an incumbent operator dominates the market on the basis of a nation-wide universal network in which it has invested tens of billions of ECUs. The key to understanding the opportunities and the structure of competition in such an environment is interconnection. By using the bottleneck facilities of the incumbent's network, new competitors can concentrate their activities on market segments where competition is viable, need not set up fully-separated stand-alone network solutions and, nevertheless, are able to offer end-to-end services to their customers on a nation-wide basis. Although there remain economies of scale in some parts of the telecommunications network, interconnection is desirable because it enables competition to work efficiently in areas where there are less economies of scale. Pure stand-alone solutions would limit the scope of competition in telecommunications significantly. The barriers to entry would be much higher when compared with a solution where interconnection of competing operators is possible.

Interconnection is not only essential from the competitor's point of view. It is of equal importance from the user's point of view. With interconnection, users get all the benefits of the integrated monopolistic system and in addition the benefits of competition. Without interconnection the benefits of competition to users would be less obvious. They would have to compare advantages of competition with disadvantages.

1.2. Interconnection as a regulatory challenge

The interconnection of networks is economically beneficial and, therefore, should be a predominant objective of any telecommunications policy. Because of the competition-supporting nature of interconnection, new market entrants generally require interconnection under fair and equal conditions when compared with incumbent telecommunications operators (TOs). In addition, because of the positive network externalities, interconnection of networks is in the interest of telecommunications users.

Given the necessity of interconnection for effective competition in telecommunications, there are reasons to assume that the dominant (if not monopoly) integrated-network operator with bottleneck control over access to most customers would regard that access as its major competitive advantage and would be less willing to provide the public-interest solution for interconnection. The incentives for the incumbent to offer interconnection on a fair and efficient basis are mixed. These incentives may be positive if the services other operators offer are complementary to its own. The contrary is true, however, if the services are substitutive and competitive. Then the incumbent may expect disadvantages from interconnection and might attempt to refuse it altogether, offer it at too high a price or at too low a quality. This incentive structure leads to our presumption that pure market solutions based on commercial negotiations for interconnection will not lead to socially-efficient outcomes. This evaluation defines the regulatory challenge with regard to interconnection. Regulatory intervention should aim to compensate for the market-failure aspects of the commercial relationship between different network operators. Like any regulatory intervention, it should conform to the market as much as possible and should make maximum use of the market mechanism.

1.3. Open network provision and a European interconnection policy

The European Union is on the way to fully opening up the telecommunications markets for competition. In June 1993, the Council decided to end definitively the voice-telephony monopoly in the member states by the end of 1997. A decision to set a similar date for the infrastructure monopoly of the TOs is in preparation. As an interim step, the unrestricted provision of network infrastructure for services which are already liberalised is under discussion. Each of these liberalisation measures will require the interconnection of new networks with existing ones to develop viable and efficient competition.

At the same time, the whole regulatory approach of open network provision (ONP) will be under review in the next few years. Although developed for and within an environment of monopolistic provision of voice telephony and network infrastructure harmonised conditions, the basic principles and essential requirements of the ONP concept are also fundamental regulatory conditions in an environment of competitive provision of voice telephony and network infrastructure. What seems to be necessary is to apply the general regulatory rules of ONP to interconnection and to develop a comprehensive interconnection policy.

1.4. *Structure of the paper*

This chapter is based on a study for the European Commission [1]. The study covers the whole variety of aspects of interconnection including technical, economic and regulatory issues and develops recommendations for a comprehensive European interconnection policy. In this chapter, we concentrate on two aspects. First, we present a systematic analysis of existing experience of interconnection in Europe and in other parts of the world. Particular emphasis is given to the regulatory regime bringing about interconnection agreements. Second, we describe and analyse the current European interconnection policy.

2. Cross-country comparison [2]

2.1. *Facts about interconnection in 16 countries compared*

2.1.1. *Different stages of evolution*

The countries analysed in our study include all member states of the European Union and the overseas countries Japan, Australia, New Zealand and the United States of America. These countries represent different levels of experience with interconnection. It is useful to classify the countries correspondingly, especially as a backdrop for the comparisons on the various issues that follow.

Basing our ranking on an overall evaluation of the experience gained by the various countries, we classify them as follows:

Much experience: USA, UK
Some experience: Japan, Australia, New Zealand
Little experience: France, Germany, Denmark, Portugal, Greece
No actual experience: Belgium, Ireland, Italy, Luxembourg, the Netherlands, Spain

The USA and the UK take a position apart, as interconnection has been part of the telecommunications markets in the two countries for ten years or more and has undergone steady further development. As regards Japan, although interconnection has been a fact there for almost as long, the development in this country has stagnated and only lately have there been adjustments and the regime put on a sound basis. In contrast, in Australia and New Zealand while the introduction of competition and, therefore, of interconnection dates back only three to four years, developments in these two countries have been so sweeping that by now some considerable experience has been gained. This justifies grouping them together with Japan in the 'some experience' category. The five countries in the 'little experience' category have been placed there as their experience has been limited to interconnection of mobile operations. None of the six countries in the last category have so far had any actual experience in the matter, although there are preparations going on in most of them.

The above discussion is reflected in Table 1. It shows for each country the realised status of interconnection for the relevant kinds of interconnection relationships (the countries are grouped according to the above classification and the groups are set off from each other by differing shadings.)

For the US and the UK, we note that they have interconnection actually realised or at least in preparation for each of the kinds of relationships considered. This indeed reflects the already very variegated structure of suppliers of telecommunications services in these two countries, not only in respect of services for end-users but also of intermediate-product type services such as those from competitive-access providers. The US is the only country that so far has had any experience of interconnection with intelligent networks.

Table 1. Status of interconnection in 16 countries

Interconnection / Country	Fixed/Fixed				Fixed/ Mobile	Mobile/ Mobile	Intelligent Network
	TO/TO	TO/Cable Operator	TO/CAP	TO/Serv. Provider (Special Network Access)			
United States	I	P	I	I	I	I	P/I
United Kingdom	I	I	I	I?	I	I	P
Japan	I	P	N	P	I	I	P
Australia	I	N	N	I	I	I	P
New Zealand	I	N	N	?	I	P	N
France	N	P	N	I	I	I	N
Germany	N	N	N	P	I	P	P
Denmark	N	N	N	?	I	N	N
Portugal	N	N	N	N	I	N	N
Greece	N	N	N	N	I	I	N
Belgium	N	N	N	P	P*	N	N
Ireland	N	N	N	N	P	N	N
Italy	N	N	N	N	P	N	N
Luxembourg	N	N	N	N	N	N	N
Netherlands	P	P	?	N	P	?	N
Spain	N	N	N	P	P	P?	N

TO Telecommunications operator CAP Competitive-access provider
I Implemented N Not under consideration
P In preparation * Agreement between TO and its subsidiary supplying mobile services

The second group of countries – Japan, Australia, New Zealand – differs from the first primarily because of the few positive entries for interconnection in the 'fixed/fixed' categories on interconnection between TOs and cable operators, TOs and competitive-access providers, TOs and service providers. The lack of positive entries for service providers, however, is also a consequence of the fuzziness of the term 'interconnection' as applied to service providers. Their use of the network is in many instances not identified as interconnection.

The countries in the third group – France, Germany, Denmark, Portugal, Greece – have in common that they all have competition in mobile communications and, therefore, corresponding interconnection arrangements but no fixed network competition. Since there is no competition in the fixed network there is also no related interconnection (with some minor exceptions in the case of France). For all these countries, changes in relevant national legislation would have to occur before competition with the incumbent in basic telephone service could be possible. As will become apparent below, although their experience with interconnection is still limited, they also contribute important insights into the terms and functioning of interconnection arrangements.

Countries in the last group – Belgium, Ireland, Italy, Luxembourg, the Netherlands, Spain – have so far not had any actual interconnection experience on which one could draw. In all but one of them, there are currently active preparations for at least the introduction of mobile competition with attendant attention to interconnection. In the Netherlands, the introduction of competition in the fixed infrastructure is also under consideration and the corresponding conditions for interconnection are under active study.

The following discussion concentrates on countries with actual interconnection arrangements in place. In the discussion on equal access (Section 2.1.4.), the focus is on the five countries where there is fixed network competition and corresponding interconnection, for which equal access is an issue.

2.1.2. *Regulatory involvement instrumental in bringing about interconnection in most countries*

We need not dwell on the fact that whenever a telecommunications market is liberalised and there are competitors with the incumbent, the newcomers seek interconnection with the established network, otherwise they would hardly be able to start business in a meaningful way. Of interest, however, is what action was considered necessary to bring about interconnection in the relevant cases. The evidence from the ten countries where there have been interconnection arrangements is shown in Table 2.

Table 2. Approaches to the realisation of interconnection

Country	Approach	Regulation: Ex ante	Regulation: Mediation/ Arbitration	Regulation: Case-by-case determination (ex post)	Competition law	Commercial agreement (no regulatory involvement)
United States	F/F	Y	Y	Y	Y	Y
	F/M	Y	?	N	Y	N
	M/M	N	N	N	?	Y
UK	F/F	Y	N	Y	N	Y
	F/M	Y	N	Y	N	Y
	M/M	N	N	N	N	Y
Japan	F/F	Y	Y	N	N	N
	F/M	Y	Y	N	N	N
	M/M	N	N	N	N	Y?
Australia	F/F	Y	Y	Y	N	N
	F/M	Y	Y	Y	N	N
	M/M	N	N	N	N	Y
New Zealand	F/F	N	N	N	Y	Y
	F/M	N	N	N	N	Y
France	F/M	Y	N	Y	N	N
Germany	F/M	Y	N	Y	N	N
Denmark	F/M	N	Y	N	N	N
Portugal	F/M	Y	N	N	N	N
Greece	F/M	Y	Y	N	N	N
	M/M	N	N	N	N	Y

Y Yes
N No

The terms used in the table have the following meaning:

- *Ex ante*: the regulator determined essential aspects of interconnection beforehand. (In the following section it is referred to as EAR.)
- *Mediation/arbitration*: the regulator stood by as a mediator – and possibly intervened

as an arbitrator if things did not move ahead – but in essence the parties negotiated interconnection on their own. (MA)

- *Case-by-case determination (ex post)*: the regulator intervened in the specific case after negotiations failed. (EPD)
- *Competition law*: there was no involvement of a regulatory agency but the application of competition law brought about interconnection. (CL)
- *Commercial agreement (no regulatory involvement)*: the parties negotiated interconnection completely on their own. There was no involvement of a regulatory authority nor the need for judicial action to enforce competition law. (CA)

We leave the detailed examination of the many entries in the table to the reader. Some summary observations, however, are of particular interest. Ex-ante determination by the regulator was used in four countries with respect to interconnection between fixed networks and in eight countries with respect to fixed/mobile interconnection. This does not preclude that in some of the same countries the other approaches have been used as well, ie, with respect to other cases. In fact, in the US all five categories identified above have been applied in at least one case. (This again testifies to the particularly diverse competitive environment in this country.) We observe that mediation/arbitration has been used in three countries with respect to fixed/fixed and in four countries with respect to fixed/mobile interconnection. With seven occurrences we also note a relatively frequent occurrence of case-by-case determination (ex post) by the regulator. Thus, for fixed/fixed and fixed/mobile interconnection, regulatory intervention of some kind can be regarded as the predominant influence in bringing about the required arrangements.

This is different in the cases of mobile/mobile interconnection. The five such cases that we observed are the outcome of purely commercial negotiations. It should be realised, however, that they have less impact on the development of competition than the ones involving fixed networks.

New Zealand is the one important exception in the sense that it is the only country where as a matter of explicit government policy, interconnection is left to negotiations between the parties subject only, if negotiations fail, to competition law as applied by the courts. As shown in the table, agreements have in fact been reached there just by way of negotiations but there have also been cases that were brought before the courts.

Table 3. Approaches to various interconnection issues (fixed/fixed)

Issue / Country	Right to inter-connect	Equal access	Points of intercon-nection	Quality of service	Principles for intercon. charges	Intercon-nection charges	Num-bering	Stan-dards	Collo-cation
United States	EAR/CL	EAR/CL	EAR/CL	EAR/CA	EAR/CL	EAR	CA/MA?	CA /EAR	EAR /CA
United Kingdom	EAR	EAR	EAR/EPD /CA	EAR/CA	EAR	CA/EPD	EAR	MA	EPD /NA*
Japan	EAR	MA	EAR	CA	EAR	MA	EAR	CA	NA
Australia	EAR	EAR	EAR/CA /MA	CA	EAR	EAR/CA /MA	EAR	CA	EAR?
New Zealand	CL	CA/CL	CA/CL	CA	CL	CA/CL	CA/MA /CL	CA	NA

EAR	Ex-ante regulation	CL	Competition law
MA	Meditation/arbitration	CA	Commercial agreement (no regulatory involvement)
EPD	Ex-post determination	NA	Not available

* Collocation was part of OFTEL's determination on Mercury's interconnection in 1985. In the 1993 determination, collocation was not addressed, and it is currently not used.

Table 4. Approaches to various interconnection issues (fixed/mobile)

Issue / Country	Right to inter-connect	Points of intercon-nection	Quality of service	Principles for intercon. charges	Intercon-nection charges	Num-bering	Standards	Billing and charging
United States	EAR/CL	EAR/CL	MA/CA	EAR/CA	EAR/CA	CA/MA	CA/EAR?	?
United Kingdom	EAR	EAR/CA	EAR/CA	EAR	CA/EPD	EAR	MA	CA
Japan	EAR	EAR	CA	CA	CA	EAR	CA	CA
Australia	EAR	EAR?	CA	EAR	CA	EAR	CA	CA?
New Zealand	CL	CA/CL	CA	CL	CA/CL	MA	CA	CA
France	EAR	EAR/CA	EAR/CA	EAR	CA/EPD	MA	EAR	CA
Germany	EAR	EAR/CA	CA	EAR	EPD	EAR	CA	CA
Denmark	EAR	CA	CA	NA	CA	EAR	EAR	CA
Portugal	EAR	EAR	EAR	EAR	EAR/MA	EAR	CA	EAR/CA
Greece	EAR	MA	MA	EAR	EAR	EAR	EAR	EAR

EAR Ex-ante regulation CL Competition law
MA Meditation/arbitration CA Commercial agreement (no regulatory involvement)
EPD Ex-post determination NA Not applicable

Table 5. Frequency of use of different approaches

	Fixed/Fixed (%)	Fixed/Mobile (%)
EAR	39	44
CA	30	39
CL	15	6
MA	11	8
EPD	5	3

2.1.3. Ex-ante regulation also predominantly applied to individual interconnection issues

Besides asking by what kind of regulatory action interconnection was brought about in general, it is of interest to ask by what kinds of different approaches the individual issues surrounding interconnection have been handled. Tables 3 and 4 address this question, the first one for the cases of fixed/fixed interconnection with respect to nine issues and the second for the cases of fixed/mobile interconnection with respect to eight issues. Table 5 provides summary statistics (percentages apply each time to the total number of occurrences which is greater than the number of cells because of double and triple entries).

We observe that ex-ante regulatory intervention (EAR) is by far the most frequent approach – for 'fixed/fixed' it applies in 39% and for 'fixed/mobile' in 44% of cases (where we have not bothered to eliminate from the total numbers the entries for New Zealand where the EAR approach is ruled out as a matter of principle). The next most frequent approach is the one by commercial arrangement (CA) with 30% and 39%. The application of commercial law (CL) is with 15% for the fixed/fixed cases of some importance (primarily because of New Zealand). The approach relying on mediation/arbitration (MA) has, with 11% and 9%, some importance, while ex-post determinations (EPD) had to be resorted to only in a few cases.

The issues to which EAR appears most often to be applied to are:

- for fixed/fixed and fixed/mobile: right to interconnection, points of interconnection, and principles for interconnection charges;
- for fixed/fixed in addition: equal access; and
- for fixed/mobile in addition: numbering;

while the issues for which CA is most often observed are:

- for fixed/fixed and fixed/mobile: quality of service and standards; and
- for fixed/mobile in addition: billing and charging.

Of some interest is that EAR applies to issues of standardisation in four out of the ten fixed/mobile cases (because of GSM) but only once in the five fixed/fixed cases. Here, reliance is placed predominantly on commercial arrangements.

Finally, we look at the frequency with which the different countries rely on the EAR approach for dealing with the various interconnection issues. The US, the UK and Australia apply it in the majority of cases while Japan, somewhat surprisingly, appears to use it relatively infrequently. It is remarkable that in two of the countries with less experience so far in interconnection matters, ie, Portugal and Greece, the EAR approach has been applied to the majority of issues, to six and seven respectively out of the eight considered.

2.1.4. Equal access as a question of costs and benefits in different environments

Equal access has been considered an essential concept in all five countries with fixed network competition. As a regulatory concept, equal access is defined from the customer's point of view. In the predominant interconnection case, that of access to the local network, equal access gives customers the opportunity to use without any bias ('equally') the trunk carrier of their choice. The choice of trunk carrier may be made call-by-call, on a more permanent basis or by a combination of the two. Besides these general aspects of customer choice, equal access requires the same type of access, the same quality and the same price.

The practice of interconnection so far shows that access to the existing telephone network was initially provided in a way that was different from the equal terms described above. This resulted from the fact that traditional telephone networks were not designed for a multi-carrier environment. Equal access cannot be incorporated into an existing network without costs and transition periods. Therefore, a decision has to be reached that the benefits outweigh the costs. Reaching such a decision may take time, and there is the need for a transition period during which networks are conditioned to meet the requirement.

Table 6 shows the time paths taken by the implementation process for equal access in the five countries. In three countries – the US, the UK and Japan – it was more than five years after the opening up of the market before regulators required that the incumbent(s) provide equal access. In the US, it was actually ordered by the Modified Final Judgment which divested AT&T in 1984. It took then several more years for the implementation actually to get under way. Consider, in contrast, Australia, where the new entrant was assured equal access as part of entry conditions and where the incumbent was given inducements to provide it quickly, which then also happened. The most revealing contrast is provided by New Zealand, where implementation of equal access was got under way five years after the opening up of the market and only three years after the competitor started business. This case demonstrates that equal access is valued by the competitor more than the price paid for it; here a price was arrived at by negotiations without the benefit of regulatory intervention (although it required arbitration by an independent industry arbitrator).

Table 6. Equal access

	Timing		Realisation period	Arrangement prior to equal access	Preselection by ballot
	Competition introduced	Regulatory requirement of equal access			
United States	~1974	1982	1984–88	CIC/PIC	Yes
United Kingdom	~1983	1991	1995–	CIC/ANI/ Blue Button	No
Japan	1985	1990	1994–97	CIC/PIC/ANI	Open
Australia	1992	1991	1992–1995	CIC/ANI	Yes
New Zealand	1989	NA*	1994–	CIC/ANI	No

NA* Not applicable. Equal access was the outcome of negotiations, finally decided through arbitration under court supervision.
CIC Carrier-identification code
PIC Personal-identification code
ANI Automatic-number identification

Interconnection practice at the beginning of network competition allows for several forms of 'unequal access' with more or less inconvenience to the customer and more or less competitive disadvantage to the new entrants. Table 6 shows that in the five countries the most common approach has been the use of a carrier identification code (CIC) in combination with a personal identification code (PIC) or automatic number identification (ANI). In the UK, the inconvenience of unequal access for Mercury customers has partially been compensated for on the terminal-equipment side. Mercury customers usually use a Mercury telephone handset where they simply have to press the 'blue button' and only need to dial the normal telephone number for making phone calls. The handset compensates for the inconvenience of dialling the CIC and the PIC. This solution comes close to equal access from the customer's point of view. From the carrier's point of view, this evaluation depends on the question of who has to pay for the added equipment costs of this type of access.

2.1.5. Different approaches to financing universal-service obligations

Table 7 has been constructed to throw some light on the relationship between universal-service obligations (USOs), a universal-service fund (USF), and access charges [3].

We observe from the table that for all ten countries, regulation imposes USOs on carriers. They consist of the requirement either to cross-subsidise particular services through an unbalanced tariff structure or to fulfil some other obligations. Only Denmark has a balanced tariff structure.

A USF exists in two countries, the US and Australia. In these countries access charges have, however, also been imposed so that competitors contribute to the cost of USOs. In other words, the USF is not considered to be sufficient to cover all relevant costs. It should be noted, however, that an alternative to access charges is in place in these countries that would at least in principle allow that USOs be completely financed by this route.

On access charges independent of the existence of a USF, we note that they are in use in all countries with fixed network competition or are at least under consideration. We make the distinction between explicit and implicit use of this instrument. In Australia, the regulator would insist that there is no access charge; the existence of the Australian CAN charge element in interconnection charges is, in our view, the implicit imposition of such a charge under a different name.

Table 7. Universal-service obligations and access charges

Country	Universal-service obligation			Universal-service fund	Access charges			
	Tariff structure		Other USO		fixed/fixed		fixed/mobile	
	In balance?	Rebalancing ?			explicit	implicit	explicit	implicit
United States	N	Y	Y	Y	Y	Y	N	N*
United Kingdom	N	Y	Y	N	Y	N	Y**	N
Japan	N	Y	Y	N	Y	N	N	N
Australia	N	Y	Y	Y	N	Y	N	?
New Zealand	Y?	Y?	Y	N	Open	N	N	Y
France	N	Y	Y	N	NA	NA	Y	N
Germany	N	Y?	Y	N	NA	NA	N	Y
Denmark	Y	N	Y	N	NA	NA	N	N
Portugal	N	Y	Y	N	NA	NA	N	N
Greece	N	Y?	Y	N	NA	NA	N	?

NA Not applicable
Y Yes
N No
* Possibly in some states
** In principle, but currently waived

Only in three of the ten countries are mobile network operators required to actually pay an access charge component; in two of them it is implicitly included in the interconnection charge. The most interesting instance of this is in the German case where the very method of computing the interconnection charge assured that the access charge was included without even making reference to it (see Section 2.1.8.). Access charges as an explicit measure for mobile operators exist in the UK but are currently waived. Therefore, according to a recent decision in an arbitration proceedings, France is the only country where mobile operators actually have to pay an explicit access charge.

2.1.6. Public accessibility of interconnection agreements

The issue of transparency of interconnection agreements appears to be handled differently according to whether there is fixed network or only mobile network competition. As we observe in Table 8, in four of the countries with fixed/fixed interconnection there is public access to at least the regulated part of interconnection agreements. The exception is Japan, where only the regulator has access to the agreements. In the countries with only fixed/mobile interconnection, in four cases only the regulator has access to the agreements while in one case, Portugal, the regulated parts of agreements are open to public scrutiny.

The following are some observations on approach and practice in the various countries:

- In the US, all interconnection agreements subject to regulatory review are on file with the relevant regulatory commissions and are open to the public. This is in agreement with the general approach to matters of regulation in this country. Nowhere else in the world are data on regulated firms open to this degree of public scrutiny, and in particular to the scrutiny of competitors, as in the US.
- In Australia, the policy appears to be related to requirements of the Trade Practices Act. Interconnection agreements in Australia can only be signed between operators of telecommunications networks with a particular licence. These agreements violate certain stipulations in the Act on non-discrimination, which therefore have to be

waived to some extent. Opening such agreements to the public permits everybody interested to verify that the waiver has been applied correctly.

Table 8. Access to interconnection agreements

Country	None	Only to NRA	Public access: only to regulated part	Public access: full access
United States			X	
United Kingdom			X	X
Japan		X		
Australia				X*
New Zealand				X*
France		X**		
Germany		X		
Denmark		X		
Portugal			X	
Greece	X***	X***		

* Except certain parts declared confidential
** The regulator may give access to certain parts of the agreements to interested operators
*** The regulator has access to the fixed/mobile but not to the mobile/mobile interconnection agreements

- In New Zealand, publication of interconnection agreements is required for the express purpose of helping to prevent collusion between the parties to network interconnection. This requirement is one of the very few specific requirements placed on telecommunications network operators.
- In the UK, the regulator has recently instituted the policy of making it mandatory to publish all new interconnection agreements that involve the dominant TO. The explicit intention is to allow every interested party to verify what interconnection terms any other party is getting, and to ensure thereby that no undue preferences are given.
- In France, interconnection agreements are on file with the regulator. The regulator is explicitly required to intervene if the agreements contain conditions that are against the public interest. Furthermore, he may give access to certain parts of the agreements to interested operators, especially when these operators have applied for licences.
- There has so far been no explicit policy statement on whether interconnection agreements belong to the private or public domain in Germany. It may be that, if Mannesmann and Telekom had been able to reach agreement on interconnection without the involvement of the regulator, as the licence actually presupposes, these terms would not have been made public. As it turned out, the charges determined by the regulator were ultimately published, precisely to inform potential bidders for another mobile licence about the interconnection terms on which to base their calculations.

2.1.7. *Evolving cost standards*

There is a regulatory requirement on the use of a particular cost methodology in four of the five countries with fixed network competition. The exception is New Zealand. In the countries that so far have only mobile network competition, three – France, Germany and Portugal – have a similar requirement. This is shown in Table 9. It appears that the closer competition gets to the core of telecommunications services, which are the services over the fixed network, the more the regulator sees the need to impose a requirement on cost standards.

The fully-distributed cost (FDC) standard is the methodology that the regulator most often requires. The reason for the preference for FDC, despite its demonstrated drawbacks, must probably be seen in tradition and in the fact that regulators have not had experience with other approaches.

One should note, however, that in some countries alternative cost standards have already been applied or are under consideration. In Australia, which in this respect is the remarkable exception, the regulator followed a directly-attributable incremental-cost (DAIC) approach when determining initial access charges for the new carrier. In the UK, for the determination of access-deficit contributions, use has been made of benchmarking where the efficiency of the regulated firm is compared with network operators abroad (the efficiency of the US Regional Bell Operating Companies served as a benchmark). Also on the regulator's agenda in the UK is consideration of moving the regulated carrier to an incremental-cost methodology. This is to some extent also true for Germany. The most important competitor in the UK, Mercury, has on its own adopted the approach of activity-based costing (ABC) which comes close to generating long-run incremental-cost (LRIC) measures. The use of price caps to control tariffs and interconnection charges, for example in the US, also lessens the distortionary impact of FDC as under this regime the regulated firm is allowed to set tariffs and charges with relative flexibility.

Table 9. Cost standards prescribed by regulator

Country	Regulatory requirements	What rule?
United States	Y	FDC + Accounting Separation
United Kingdom	Y	FDC + Accounting Separation Benchmarking
Japan	Y	FDC + Accounting Separation
Australia	Y	DAIC, FDC
New Zealand	N	
France	Y	FDC
Germany	Y	FDC, IC*
Denmark	N	
Portugal	Y	FDC
Greece	N	

FDC Fully-distributed costs
DAIC Directly-attributable incremental costs
IC Incremental costs
* The requirement is for FDC. The regulator, however, has entered into a discussion with the regulated carrier regarding a methodology consistent with LRIC

On the other hand, one should also note that for the countries having had little or no experience with competition in the telecommunications sector, the question of the cost standard on which to base the calculation of tariffs, and hence interconnection charges, is an issue which so far has attracted little attention.

2.1.8. *Relationship between local interconnection charges and local call tariffs*

Here, we attempt to get an indication of the extent to which considerations reflected in the Baumol/Willig efficient component-pricing rule for interconnection are evident in the access charges used in the ten countries. For this purpose we have sought to establish ratios of the charges to the relevant end-user tariffs. This can be done approximately for the relationship between the charges for interconnection with the local network and end-user tariffs for local calls [4]. Table 10 presents the results that we obtained this way.

It is possible to provide the following commentary on the percentages shown in the table:

- For six of the ten countries, ie, the US, the UK, Australia (initial charges), France, Germany and Portugal, the charge for interconnection with the local network is lower than the retail tariff for making a local call. We note from the third column of the table that in each of these cases the regulator was actively involved in determining the charges. One should expect the regulator to aim for this result as interconnection with the local network tends to use less resources than a local call [5]. We also note two cases, ie, Japan and Denmark, where the opposite is true. Here, however, only in the first case was the regulator involved, in the second the result was the outcome of commercial negotiations.

Table 10. Relationship between local interconnection charges and local call tariffs [6]

Country	Approximate per cent of nominal tariff	Charges determined by	Component for access charges	
			explicit	implicit
United States	~ 50%	EAR	Y	Y
UK Mercury/BT	~ 55%	EPD	N	N
Japan	130%	EAR	Y	N
Australia				
Initial	< 50%	EAR	N	Y
Subsequent	~55%	MA	N	Y
New Zealand				
Long distance	~85%	CA	N	N
Local calling	?	CL	Open	Open
Mobile	100% +	CA	N	Y
France	~70%	EPD	Y	N
Germany	82%	EPD	N	Y
Denmark	~120%	CA	N	N
Portugal	65%	MA	N	N
Greece	?	EAR	N	?

EAR	Ex-ante regulation	EPD	Ex-post determination
CL	Competition law	CA	Commercial agreement
MA	Mediation/arbitration		
Y	Yes	N	No

- The particularly low ratio shown for Australia under the 'initial' heading may reflect the regulator's use of an incremental-cost methodology in the determination of initial interconnection charges. This may change with the entrant's achievement of threshold market shares when the initial charges will cease to apply and new, 'subsequent' charges will need to be negotiated on a commercial basis.
- The entries for New Zealand reflect a complex situation. The 'long distance' entry corresponds to the charge paid by the new fixed-network competitor, Clear, for its interconnection for long-distance calls. The corresponding agreement had been concluded with Telecom in free negotiations at a time when Telecom did not yet insist on an access charge according to the efficient component-pricing rule (Baumol/Willig). The question mark in the 'local calling' entry reflects the fact that, so far (July 1994), no interconnection charge has been agreed upon for Clear's offer of local calling. For this, Telecom has demanded access charges which Clear is not prepared to accept. The case is under final review by the Privy Council in London. The 'mobile' entry reflects the interconnection agreement between BellSouth and Telecom concluded at a time when it seemed, after the first court ruling in the local calling case, that Telecom would be successful with its demand for an access charge.
- The method of determining the interconnection charge in the German case perhaps resembles most closely the Baumol/Willig prescription. The charge was arrived at by starting from the actual local call tariff and deducting from this the cost savings attributed to the particular way in which the interconnecting operator interconnects with the PSTN. It follows that the charge determined contains all the elements that are included in the tariff for the purpose of financing infrastructural and other obligations. This would at least in part explain the high value of the ratio shown in the table.

With regard to any apparent dependence of the level of the interconnection charge on an access charge component included in it, we note (looking at the fourth and fifth columns in which it is shown whether there is such a component, either explicitly or implicitly) that there is none. We observe cases in which there is indeed a relatively high level when there is an access charge (Japan, Germany, New Zealand 'mobile') but also when there is none (Denmark, New Zealand 'long distance'), and we observe a case of a low level when indeed there is no such charge (the UK) but also when there is one (the US, Australia). One is led to conclude that the inclusion or exclusion of an access charge may not be so relevant in determining the actual level of the interconnection charge.

The actual pricing methods for interconnection charges vary between countries in their sophistication and in their relationship to retail tariffs. In nearly all cases the largest component of interconnection charges is on a per-minute basis. Usually, only the costs of actually establishing interconnections are not priced on a usage basis. Peak-load pricing of usage is rare (UK), and so is pricing of call attempts (Australia).

Where interconnection charges are to be cost-based, regulators and incumbent TOs seem to have some preference for fully-distributed cost pricing of interconnection while interconnectors favour direct costing (no overheads) or incremental costing. However, price caps for interconnection services with the possibility of rebalancing are perceived by TOs as a desirable means of moving away from fully-distributed pricing principles, while they are viewed by interconnectors as a potential source for predatory TO behaviour (US).

2.2. *Lessons from the international comparison*

In all countries that have introduced network competition in telecommunications, interconnection has been among the most pressing issues. As experienced in the US from the early parts of this century onward, incumbent TOs may simply want to refuse interconnection with competing network operators. Later experiences, for example in the UK and Germany, show that incumbent TOs, even if forced to interconnect by licence or statute, do not voluntarily grant interconnection to competing network operators at terms and conditions that would be acceptable to regulators. Without regulatory intervention, negotiations often reach an impasse. This experience suggests:

a) that a right or duty to interconnect is necessary to ensure network entry;

b) that some form of regulatory intervention in private negotiations on interconnection agreements is helpful in reaching outcomes that regulators desire.

The experience also suggests that network interconnection is too complicated and possibly too individualised an issue to be regulated in all its aspects. Thus, a balance between regulation and commercial agreements needs to be reached. Since most interconnection regimes start out with a highly asymmetric market position between the incumbent TO and the interconnector, ex-post regulation has proven to be an effective tool in helping the interconnector's negotiating position. In order to create reasonable expectations about the outcome of ex-post regulatory interference, ex-ante regulation has to establish a set of basic guidelines for the issues that can be treated through ex-post regulatory determination and the criteria to be used by the regulator. The UK framework has created such rules in great detail. It is questionable whether the UK level of detail is needed or whether more general rules would suffice. The trade-off is between constraining the regulator and requiring frequent changes in the ex-ante regulation on the one hand and regulatory discretion and vagueness on the other.

Most countries use their interconnection regime as an active regulatory tool for promoting competition in the telecommunications sector. The objectives of interconnection regulation as revealed by regulators in our country studies include entry help and the viability of the incumbent. Entry help is provided to the interconnector via low interconnection charges (or high prices for retail telecommunications services or both) that improve their competitive position via the incumbent TO. Entry help is also provided via rules that equalise interconnection charges across interconnectors and thereby improve the competitive position of small interconnectors relative to large interconnectors (US 'equal charge' rule). As appropriate, entry help is most pronounced at the beginning of the interconnection regime (US, UK). While the question of the viability of the incumbent TO is raised at the outset, it becomes a more pressing issue over time, as 'bypass' occurs and the incumbent loses market share. A major reason given for this problem by incumbents is the slow speed allowed to rebalance retail rates.

The viability of the incumbent is taken care of through fully-distributed cost pricing of interconnection and through (implicit or explicit) access charges or access-deficit contributions. We have not found any satisfactory regime of access charges among the countries we studied. The US is actively searching for a replacement of the status-quo approach currently used (which was derived largely from a formula for separating interstate and intrastate costs). The UK approach is much more explicit and rational than the US approach but it requires superhuman objectivity in making waiver decisions as well as complicated and questionable calculations. Access charges have been justified in the various countries by constraints on tariff rebalancing or by USOs or both. While constraints on tariff rebalancing should be lifted to accommodate competition and to promote cost-based pricing, USOs may actually become more desirable under competition. So far, only Australia has a method for establishing and financing the costs of USOs in place. Experience in this area is needed.

The interaction between regulatory bodies and official competition policy in the area of interconnection varies among the countries studied. In New Zealand, competition policy fully dominates; there is no regulation. In the US the two types of policy often compete with each other. Major parts of the US interconnection regime are the result of competition policy, for example, the vertical separation into local and long-distance carriers, the equal-access stipulations and the equal-charge rule. In the UK interconnection regime, the regulator seems to actively pursue competition policy, without facing strong constraints from other instruments of competition policy. Our expectation is that active regulation of interconnection will eventually disappear as the market for interconnection matures. Competition policy will then have to take over as the main policy tool for dealing with problems arising in this market. It is therefore imperative that the role of competition policy be established early on.

Most countries have gained their first experience with interconnection agreements through fixed-mobile interconnections. For fully-fledged network competition, fixed-fixed interconnections are currently more important. The question is, to what extent can countries learn from their fixed-mobile interconnection experience for the forthcoming fixed-fixed case and to what extent they are locked in by their decisions on fixed-mobile interconnections. For several countries a particular emphasis of fixed-mobile interconnection agreements is on low lease rates for private circuits. In the UK, as the only European country in our study with fixed-fixed interconnections, preferential rates for private circuits have not been provided under fixed-fixed interconnection agreements. The ostensible reason for this asymmetry has been to encourage facilities investments by competing fixed-network operators and to discourage similar investments by mobile operators. That there can be ample room for private circuits in fixed-fixed interconnection agreements has been demonstrated by the importance of the US special-access regime.

Over time, interconnection moves from something infeasible or difficult to something resembling business as usual. At the beginning of interconnection regimes, there is usually a single entrant that wants to be interconnected to the incumbent TO. The timing to realise interconnection along with starting interconnection charges appears to be the most pressing issue. While timing to realise interconnection remains important for the newcomers, the feasibility of interconnection has long been established, and most parties face other issues. Countries with long experience in network competition (US, UK) are confronted with a diversity of parties that want to interconnect and that have different interests. An interconnection regime that individually takes care of these interests is likely to become cumbersome and complex. Rather, standardisation of agreements on many issues appears possible, leaving fewer issues that have to be solved for individual cases. As the market for interconnection matures, non-discrimination between interconnectors becomes pressing, and the case is made that all potential interconnectors should be treated similarly (as in US expanded interconnection).

How can countries at the beginning of their interconnection regimes learn from the more advanced countries? Clearly, the more advanced countries have gone through stages in their interconnection regimes that might be skipped or shortened by the followers. For example, the technical feasibility of interconnection has by now been demonstrated under various network configurations. There is ample experience with location of interconnection points and with compatibility between different types of networks and line capacities. Similarly, there is a long list of issues that need to be considered for interconnection agreements where the solutions could be adapted from UK interconnection agreements or US interconnection tariffs, both in the public domain. Some issues are not so clear cut but nevertheless learning is likely to be helpful. For example, the costs and benefits of equal access, while possibly different in magnitude across countries, are unlikely to differ much in kind. Methods for cost-benefit analysis of equal access may therefore be transferable.

In the US and the UK, the publicity of the contents of interconnection agreements goes along with an involvement of the whole telecommunications sector in the process that shapes the interconnection regime. Such an involvement and public discussion may slow down the speed of decision-making but it is likely to improve the understanding of the issues and trust in the outcome of regulatory decisions.

The US country study is the only one that could be relevant to the relationship between the European Commission and the member states. The US experience suggests that dual regulation causes problems of jurisdictional cost separation and of conflicts between rules that govern intrastate and interstate transactions. The US has generated reasonable uniformity of approaches throughout the country where either the federal regulator has preempted state regulation or where competing interstate/intrastate regulation exists and operators can arbitrage between jurisdictions. Translated to the case of the European Union, this suggests that the European Commission could either prescribe binding rules on interconnection or establish European standard rules that can be followed on a voluntary basis. Voluntary rules could lead to a unified approach to the extent that arbitrage possibilities for telecommunications services between countries can be realised through facilities-based competition through interconnection.

3. Aspects of a European interconnection policy

3.1. Introduction

All experience with competition in the European Union and in the rest of the world proves that interconnection is the key for transforming the former monopolistic market structure in telecommunications into a competitive one. Viable competition is unthinkable without interconnection between mainly (but not exclusively) the incumbent's and the new competitors' networks. Economic analysis can show and can give the answer to why that is the case. Technical analysis gives evidence that interconnection is feasible and more cost-effective than stand-alone solutions. Not only competition in public voice telephony requires interconnection of networks. The status of service competition in the EU proves that viable service competition in other service fields benefits from interconnection to existing networks.

Given the (initial) market structures that exist before viable competition emerges in the telecommunications sector, the supplier(s) and user(s) of interconnection services will not produce results that make best use of the competitive potential in the market. Market-structure asymmetries give incentives to the dominant supplier not to offer the degree and the conditions of interconnection which best serve the public interest. This result can be derived from theoretical economic analysis. Our rich empirical case studies show that this result obviously describes most of the worldwide experience with interconnection in the real world. The case for policy and regulatory involvement can be made quite easily. This holds on the national as well as on the European level.

3.2. Arguments in favour of a European interconnection policy

3.2.1. Liberalisation on the European level

The European Union took the lead in transforming the telecommunications sector from a monopolistically-structured market to a competitive one in the eighties. Many (or even most) of the member states have adopted their national regulatory framework according to the requirements of the services directive [7] not earlier than required by the directive but in line with it. They have accepted and are willing to accept policy formulation on major telecommunications-policy issues at a European level. These member states expect that policy frameworks and measures which govern and structure fundamental policy decisions will also be developed on the European level.

Given the fundamental principles of the Treaty [8], the interaction between European telecommunications legislation and national telecommunications policy as described above does not prevent any member state from opening its telecommunications markets to competition earlier than required by European legislation.

In July 1993, the European Council adopted a resolution [9] on the further development of liberalisation in the European Union. Liberalisation of all voice-telephony services by and not later than 1 January 1998 is the intended policy goal. In this context, the development of a framework for interconnection agreements and the definition of principles for access charges has been regarded as a line of action which has to be worked out.

In May 1994, the Bangemann group at the request of the European Council made far-reaching recommendations on the future liberalisation path in Europe [10]. The report recommends acceleration of the ongoing process of liberalisation and the opening up of the infrastructure and services to competition as soon as possible. Furthermore, the report recommends interconnection of networks and interoperability of services as primary Union objectives. On this basis and other contributions, European decisions on a fully-liberalised telecommunications market can be expected soon.

Given this status of policy decisions, the member states expect a framework for interconnection or a European interconnection policy to be developed in the near future.

3.2.2. Divergence in national approaches

A common European interconnection policy is easier to implement when member states have not yet dealt with interconnection in their national regulatory framework. This statement can also be formulated the other way around. When all or most member states have developed a national interconnection policy and have realised interconnection agreements between networks operators, and the approaches chosen vary a lot, the question on the usefulness (or the transaction costs) of a European interconnection policy is more difficult to answer. In any case, if many interconnection agreements are in place a new interconnection policy requiring amendments or changes to existing agreements is difficult to implement.

Where do we stand in Europe on national interconnection policies and interconnection agreements? The situation differs with market segments or interconnection scenarios and is, of course, dependent on the steps towards competition which have already been taken in the member states.

Overall, the situation in the various member states is heterogeneous. Some member states have already developed (elements of) an interconnection policy. In most member states, however, no substantial steps towards developing a regulatory policy for the interconnection of networks and services have so far been taken, and many legal, institutional, technical and economic barriers against (viable) interconnection still persist.

On interconnection between fixed networks, only the UK has experience with that type of competition and, therefore, those types of interconnection agreements. As the UK country study shows, interconnection evolves to become such a complex issue, that policy development in this field is not or need not necessarily be a one-step decision-making process. In some other member states the first ideas and concepts to emerge were on how to introduce competition into the core telecommunications business (network infrastructure, voice telephony). In these countries the first ideas and elements towards an interconnection policy for fixed networks are just emerging. A European interconnection policy could fill a gap for fixed-network interconnection.

The situation is different with mobile networks. Most member states have introduced or are introducing competition in mobile telephony. Therefore, they had to develop an interconnection framework or are doing that currently. Several interconnection agreements have been negotiated within these frameworks. The regulatory frameworks and the agreements themselves differ in a variety of details and outcomes. The mobile carriers face quite different points of departure in their competitive position between member states. A lot of restrictions still prevent them from providing the most efficient mobile networks and services. Many barriers to efficient interconnection still exist for mobile operators, in particular in direct connections between mobile networks in different member states. A European interconnection policy could harmonise interconnection frameworks and conditions.

Together with developing ONP conditions for the telephone network and setting conditions for access to its (advanced) features, some member states are taking the first steps towards interconnection in an intelligent-network environment. These approaches are still at an early stage in most member states. A European policy in this field could take a lead for innovative and uniform approaches.

3.2.3. *European-wide networks and services*

The main (but not the only) argument in favour of a comprehensive European interconnection policy consisting of similar conditions for access to networks and network functions under equal terms and conditions in each member state can be made with the requirements of European-wide networks and services. If network operators or service providers want to run a trans-European network or want to offer a European-wide service, they will have to rely on interconnection to national public networks in many cases. Interconnection can be a bottleneck if it is not provided everywhere. It can also be a bottleneck if the technical, economic and regulatory terms and conditions vary significantly between member states.

This argument as the basis for a European interconnection policy received the highest legal support in the Maastricht Treaty on European Union. The Treaty introduced a new 'Title XII' on 'trans-European networks' into the EC Treaty and thereby defined a new field of competence for the European Union. Article 129b sets the legal principles of an interconnection policy on trans-European networks: 'The Community shall aim at promoting the interconnection and interoperability of national networks as well as access to such networks'. The basic policy on interconnection now has a direct legal basis in the constitution of the European Union.

3.2.4. *Harmonising national interconnection policies*

As we will show in Section 3.3, the European Union has developed an initial framework for a European interconnection policy, mainly in the proposed voice-telephony directive. Compared with the regulatory and policy frameworks in countries that already have some years of experience of competition and interconnection, the European policy framework as defined so far is consistent with a variety of very different interconnection policies in the member states. A harmonised approach in the European Union would require a much more detailed approach if that is the intended objective in Europe.

3.3. *Current European interconnection policy*

3.3.1. *The open network provision framework*

Basic principles of an European interconnection policy have been developed within the ONP framework. When the ONP concept was born in the eighties, service liberalisation and the provision of transborder service offerings was at the centre of the Commission's policy approach. The network-infrastructure monopoly of the TOs was generally accepted at that time in Europe. Furthermore, most member states wanted to reserve certain basic services for exclusive provision by their TOs.

In that environment the ONP concept was born: TOs should be obliged to open their network under fair competitive conditions to competing service providers and to interconnect with them. Working out the principles of the provision of the network was regarded as a necessary requirement for a Union-wide competitive market. One of the main reasons for developing the ONP concept was to avoid a series of contentious cases and lengthy conflicts about infringements of the competition rules of the Treaty. Besides allowing access for competitive service providers, interconnection between TOs in different member states was a concern of the Commission to allow efficient Union-wide communications.

The so-called ONP framework directive [11] lays down the principles under which service providers and users can get access to reserved services and the public telecommunications infrastructure. The directive defined ONP as 'the harmonisation of conditions for open and efficient access to and use of public telecommunications networks and, where applicable, public telecommunications services'. ONP is intended to facilitate the provision of services using public networks or public services or both. According to this aim, ONP is intended to promote fair competition between vertically-integrated TOs and service providers which have to rely on the services and resources of the TOs to produce and offer their own services. The second goal of ONP is to harmonise technical interfaces, usage conditions and tariffs to facilitate the provision of pan-European telecommunications services. Harmonisation will be realised on the basis of European standards to be adopted by the European Telecommunications Standards Institute (ETSI).

Three types of principles and application criteria define the ONP regulatory framework, namely:

1) harmonised conditions,
2) basic principles,
3) essential requirements.

ONP conditions include, in particular, *harmonised conditions* on:

- technical interfaces, including the definition and implementation of network termination points;
- usage conditions, including access to frequencies;
- tariff principles.

The main emphasis with technical interfaces is their conformity with European standards. For existing services and networks, existing interfaces should be adopted. Also, new services should rely on existing interfaces, and if that is not possible, new interfaces have to be specified. For new networks, ONP requirements should be taken into account when specifying new interfaces.

Usage and supply conditions refer to a standard set of contractual conditions governing access and usage including, for example, delivery period, quality of service, maintenance, fault reporting, procedures. They may include conditions for interconnection with public and private networks, conditions for resale of capacity, and conditions for shared use.

Tariffs should mainly comply with a set of basic principles defined below. Further general requirements for tariffs are that they are cost-oriented and sufficiently unbundled. Furthermore, tariffs must be non-discriminatory and guarantee equality of treatment.

ONP conditions must comply with a number of *basic principles*, namely:

- they must be based on *objective criteria*;
- they must be *transparent* and published in an appropriate manner;
- they must guarantee *equality of access* and must be *non-discriminatory*, in accordance with Union law.

These basic principles apply to all harmonised conditions of ONP.

The general philosophy of ONP is to realise and guarantee open access to public networks and services, and to avoid restrictions on that. The directive, however, takes care of a set of 'non-economic reasons in the general interest' which give national regulatory authorities (NRAs) the right to restrict access to the public telecommunications services. Access for users and service providers can only be restricted for reasons based on the following *essential requirements*:

- security of network operations;
- maintenance of network integrity;
- interoperability of services, in justified cases;
- protection of data, as appropriate.

In applying restrictions on access, regulators have to follow the principle of proportionality and they shall not be excessive in relation to the aim pursued.

3.3.2. Current open network provision legislation and initiatives

While the framework directive defined and set the general principles of ONP, the application to specific services remains subject to separate legislative measures. In the last few years, the following areas have been addressed for legislative ONP measures:

- leased lines;
- integrated-services digital networks (ISDN);
- packet-switched data services (PSDS);
- voice telephony.

a) Leased lines. The leased-line directive [12] mainly deals with conditions and specifications on how leased lines should be provided. The directive contributes to interconnection regulation and service competition in so far as it no longer accepts technical restrictions for the interconnection of leased lines among each other or for the interconnection of leased lines and public telecommunications networks, because they can be replaced by less restrictive regulatory measures.

b) Integrated-services digital networks. Although the Council recommendation on the provision of ISDN [13] in accordance with ONP principles aims at harmonising the conditions for open and efficient access to and use of ISDN, it does not address interconnection issues very specifically. It is stated that users are requiring further access arrangements such as M- and U-type interfaces which are not yet provided by TOs. This issue still is under study and not yet settled.

c) Packet-switched data services. Similar to ISDN, this Council recommendation deals mainly with supply conditions for PSDS and not with access and interconnection issues.

3.3.3. The interconnection model of the voice-telephony directive [14]

The ONP rules mentioned above were primarily aimed at the definition of minimum sets of service features and conditions. More precise and detailed rules on interconnection have been developed in the recently adopted Council directive on the application of ONP to voice telephony.

The directive aims to 'create the best possible regulatory environment for boosting the competitive provision of pan-European services using capabilities and functions of the public switched-telephone network and service, whilst safeguarding the ... provision of universal service'. On interconnection, the regulatory principles primarily aim to avoid abuse of a dominant position by TOs. The document deals with the provision of access to the public voice-telephony service and the network over which that service is delivered for:

- end-users;
- competitive service providers;
- public mobile-telephony services operators;
- other TOs.

Access to voice telephony includes access to intelligent functions of the public network.

ONP conditions for access to the fixed public-telephone network shall apply to all the network technologies currently in use including analogue telephone networks, digital networks and the ISDN. The directive does not apply to mobile-telephony services in general, but it does apply to the use of the fixed public-telephone network by operators of public mobile services. It does not apply to the direct interconnection between operators of public mobile-telephony services.

In its interconnection provision (Art. 11), the directive seems to exclude the case for the interconnection of two fixed network operators within the same member state. The wording in clause 22 is different and seems to regard the interconnection case mentioned before to be covered by the directive. Perhaps there might be some openness in the scope of the interconnection provisions of the directive.

Given this little (but important) classification problem, the directive addresses and is applicable to the following interconnection scenarios:

1) Interconnection between two fixed public-telephone networks from different member states.
2) Interconnection between a fixed public-telephone network and a public mobile-telephony network within the same member state.
3) Interconnection between a fixed public-telephone network and a public mobile-telephony network from another member state.

This classification on the other hand means that the directive is not applicable to the following interconnection scenarios:

4) Interconnection between two mobile-telephony networks within the same member state.
(5) Interconnection between two fixed public-telephone networks within the same member state.
(6) Interconnection between mobile networks from different member states.

Clause 24 addresses scenario 5), where more than one TO operates a fixed telephone network in the sense that although not directly applicable, the interconnection arrangements 'should take due account of the principles laid down in this Directive'. If this distinction holds, there is no systematic criterion on the scope of the directive. It neither totally excludes mobile networks nor concentrates on crossborder interconnection scenarios. The outcome on applicability is more a mixture of these two criteria.

The directive basically distinguishes between three types or ways of accessing the PSTN. The regulations mainly deal with standard access, which means interfaces for access at commonly-provided network termination points. These types of access are relevant for all or at least major parts of the whole user community.

Besides commonly-provided network termination points, the directive deals with special network-access requirements, which will mainly be relevant for service providers but might also have some relevance for some groups of larger users. Finally, the directive sets regulatory rules for the access or interconnection services which other network operators demand. Figure 1 gives an overview on the three types of access which are handled in the voice-telephony directive.

According to Art. 10, users can request access to the fixed telephone network at network termination points other than those offered for 'standard access'. TOs shall respond to such requests if they are reasonable (in terms of technical feasibility and economic viability) and NRAs have to ensure that they provide such types of access. A TO can only restrict or deny that access after getting the agreement of the NRA. Service providers or (larger) users shall obtain special network access to public network resources without waiting for time-consuming standardisation procedures. The Commission can, however, initiate standards for new types of network access within the ETSI process. They can engage in technical and commercial arrangements with the TO. There is no obligation to publish such off-tariff agreements which makes it easier for TOs to discriminate. As a regulatory safeguard against such behaviour, NRAs have the right of access to such arrangements. The NRA may intervene on its own initiative and shall do so if requested by any party. Service providers, however, get no right to collocate such that they can install their equipment on the premises of a TO.

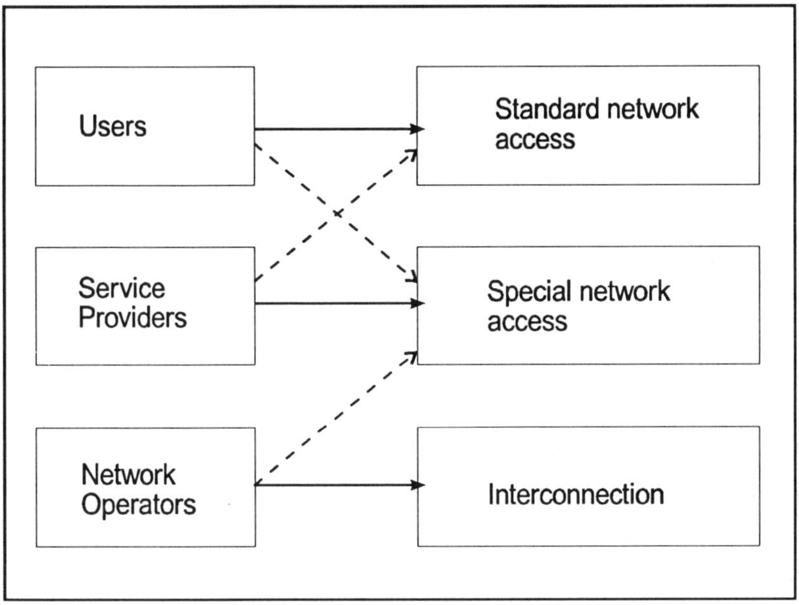

Figure 1. Types of network access

Art. 11 sets up a regulatory framework for interconnection in a rather broad way. The main focus of the directive is the Union-wide provision of voice-telephony services. Therefore, the directive only deals with the three scenarios of interconnection mentioned above.

The first *basic principle* of the regulatory framework is the right to interconnect. Operators of public mobile-telephony services in the same member state and from other member states as well as fixed public-telephone-network operators from other member states have the right to interconnect to the national fixed public-telephone network. This right to interconnect can only be restricted with prior agreement of NRAs. Restrictions might be based on essential requirements.

The *second basic principle* is that regulators should mainly rely on negotiations between the parties involved to get agreement on technical and commercial arrangements for interconnection.

The *third basic principle* is the right and duty of the regulator to intervene in interconnection matters. Interconnection 'should be subject to regulatory oversight in order to safeguard the Community-wide interests of users and ensure compliance with Community law ...'. NRAs might intervene on their own initiative and shall do so if requested by either party. The forms and types of intervention are not specified, but they include the right to set conditions on interconnection agreements. The criteria and objectives of regulatory actions are limited to conditions which are:

- non-discriminatory;
- fair and reasonable for both parties;
- offer greatest benefit to all users.

Besides meeting these criteria, NRAs also have to ensure that interconnection agreements:

- are entered into and implemented in an efficient and timely manner;
- include conditions on conformance to relevant standards;
- include conditions on conformance to essential requirements;
- include conditions on the maintenance of end-to-end quality.

It is not specified whether and to what extent such requirements should be set ex ante by the NRA or whether negotiated interconnection agreements should be controlled by their compliance with these requirements. Furthermore, there is no general approval requirement for interconnection agreements by NRAs.

In Art. 27, the directive sets up some conciliation and dispute resolution procedures which might also be used in interconnection disputes. Thereafter, unresolved disputes with a TO by any party about an alleged infringement of the provisions of the directive can be appealed to the NRA or another independent body. Easily-accessible and inexpensive procedures shall be created to resolve such disputes in a fair, timely and transparent manner. When more than one member state is concerned, the dispute may also be referred to the Commission.

The *fourth basic principle* is the possibility of access charges being included in interconnection agreements. In a complicated language they are introduced as 'specific compensation provisions [15] for the telecommunications organisation in situations where different operating conditions, e.g. price controls or universal service obligations, are imposed upon the respective parties ...'. Under this provision, many situations might potentially give reason for access charges when even the monopolistic price control is regarded as an asymmetric type of regulation which has to be compensated for. Only very broad and unspecific criteria are set for the application of these 'compensation provisions' or access charges: they shall be cost-oriented (which costs?), non-discriminatory and fully-justified (according to what objective?), and need approval of the NRA.

The *fifth basic principle* requires details of interconnection agreements to be made available to NRAs upon request. This principle has more relevance in what it does not require than in what it does require. It does not require that interconnection agreements need prior approval by the NRA before they become effective. Furthermore, it does not require public access in whole or in part to interconnection agreements.

There are no interconnection-specific tariff and cost-accounting principles defined or set, which means that only the general principles on voice telephony as a whole shall apply. These are first of all the principles of transparency and cost orientation as specified in the ONP framework directive. Tariffs should be 'sufficiently unbundled', independent of the type of application which the users implement and may include 'tariff constraints (imposed by the NRA) relating to the objectives of universal telephone-service accessibility including town and country planning'. Of some relevance for interconnection might be the provision that bulk-discount schemes can be offered (Art. 14(1)). The tariff regulations should be supported by suitable cost-accounting systems which are deemed to control tariff principles and restrictions. The cost-accounting rules are formulated according to a typical fully-distributed costing approach (Art. 13(3)).

To summarise, the voice-telephony directive favours a 'light-handed' regulatory approach towards interconnection. It does not rely on or require ex-ante regulation of interconnection agreements, but provides for regulatory intervention in case commercial negotiations on interconnection fail. The directive leaves open questions on its jurisdiction over some (relevant) interconnection scenarios and it is not very specific in its provisions on interconnection in an intelligent-network environment. Besides referring to the possibility of access charges being included in interconnection agreements, the directive does not contribute to the definition of universal-service principles and conceptual and implementation aspects of interconnection and access charges. The Commission regards the provisions for access and interconnection agreements in the directive as a sufficient initial framework at the Union level. With phased progress towards increased liberalisation, this framework may have to be reviewed.

3.3.4. Developing universal-service and access-charge principles

Although interconnection and universal-service issues are not necessarily directly interrelated issues of telecommunications policy, in practical terms and in all relevant experience around the world these issues are directly interrelated. Any coherent interconnection policy has to deal with and has to decide whether or not interconnection charges should also take care of the universal-service obligations of incumbent telecommunications carriers and how and to what extent they should include access-charge elements. A coherent policy in this field has to specify universal-service principles, has to identify those operators that have to carry the burden of universal-service obligations, if there are any, has to decide on the appropriate financial mechanism, and how or whether to share the burden with other suppliers in the industry. Only when the final point is reached, does the question of access charges arise compared with other burden-sharing principles and mechanisms.

The discussion on universal service in telecommunications has started in the European Union. The overall goal is broadly accepted in the European Union. Some basic principles have been formulated and the debate on financing universal service has started. Basic principles of universal service in telecommunications, which have already been briefly mentioned in the Council resolution on the review of the situation in the telecommunications sector in July 1993 [16], have been further developed and specified in a Council resolution of February 1994 on universal-service principles in the telecommunications sector [17].

The basic objective of the resolution is to provide 'assistance in achieving the goal of universal service in a competitive environment and to the progressive rebalancing of tariff structures ...'. Furthermore, major elements constituting universal service at Union level are to be identified. The resolution refers to the various ONP measures which are assumed to have identified the basic elements for a definition of universal service and does not give a specific definition of universal service itself.

Maintenance and development of a universal telecommunications service is regarded as a key factor for the future development of telecommunications in the Union. Universal service is not regarded as a static concept realised by a set of basic-service elements; instead, it is regarded as a dynamic concept that 'must evolve to keep pace with advances in technology, market development and changes in user demand'.

Universal service implicitly is being defined by a set of principles, namely: universality, equality and continuity. These principles are a basis for such a service to permit access to a defined minimum service of specified quality to all users everywhere and at a reasonable price. Common principles for the provision of universal service should be defined throughout the Union while taking account of specific national conditions.

On financing universal service, the resolution only refers to a basic voice-telephony service and not to other potential candidates for universal service. Under the specific circumstances where this service 'can only be provided at a loss or can only be provided under cost conditions falling outside normal commercial conditions, that service may, where justified and subject to the approval of the national regulatory authority, be financed through internal transfers, access charges or other mechanisms which take due account of the principles of transparency, non-discrimination and proportionality, while ensuring compliance with competition rules in order to make a fair contribution to the burden which the provision of universal service represents'. Thus, access charges as an instrument of financing universal service are not favoured per se. It is one instrument besides others. On the basis of this resolution, the Commission is invited to elaborate common access-charge principles.

The discussion on universal service in the European Union mainly refers to the ONP framework developed so far. The ONP practice so far has developed the following universal-service concepts:

- the basic provision of service;
- quality of service;
- tariff principles (including targeted tariff schemes for particular user groups);
- dispute-resolution mechanisms;
- special public-service features (for example, emergency services);
- certain Union-wide service features.

The Commission in a related communication considers the principles provided in the context of ONP as sufficient at this stage to form a basis for universal-service conditions.

Financing universal service is seen in the context of tariff adjustment followed by the introduction of competition in the core markets. The adjustment process should be accompanied by a number of factors:

- the costs of meeting universal service have to be identified;
- capacity should be retained to finance unavoidable deficits in local access from more profitable parts of the business;
- gradual tariff rebalancing must be permitted;
- new entrants should make an appropriate contribution to the provision of universal service and to the transfers required to finance it via a system of access charges;
- sufficient national flexibility should be given in line with the principle of subsidiarity;
- social and economic cohesion of peripheral regions should be taken into account.

Internal transfers and access charges are not seen as an alternative to tariff rebalancing and, without explaining how, access charges should not be used as a means of controlling a competitor's cost structures. Access charges should be structured in such a way that necessary improvements in cost efficiency or productivity gains are not delayed by shifting the burden of inefficiency onto new market entrants. Generally, access-charge systems should be based on the principles of transparency, non-discrimination and proportionality.

3.3.5. Interconnection positions in the Green Paper on mobile communications

In its Green Paper [18] on a common approach in the field of mobile and personal communications, which was published in April 1994, the Commission has developed a policy approach towards interconnection between different mobile networks and between fixed and mobile networks. In the Green Paper, which has the character of a policy paper, the Commission has proposed positions for legislative measures and other lines of action. The positions and recommendations of the Green Paper are now in the stage of public comments.

The Commission emphasises the vital importance of fair and efficient interconnection agreements with the fixed network for mobile operators. Payments to fixed-network operators for the conveyance of calls and the provision of leased lines account for 30–50% of total revenues of mobile-network operators and, therefore, determine to a large extent the economics of such networks. The Commission criticises restrictions on direct connections to other mobile operators, both within and between member states, which exist in some member states. Such restrictions hinder Europe-wide roaming and the establishment of trans-European networks.

Although the Commission identifies a lack in the definition of interfaces, it does not regard it as necessary to establish further specific directives on these interfaces and related interconnection conditions at a Union level. Instead, the framework provided by the ONP framework directive is regarded as sufficient, provided that interconnection is subject to strict supervision by NRAs.

To facilitate interconnection and interoperability, substantial progress is required in the definition of European standards. Standards are required for the following most essential interfaces:

- the interface between mobile networks and the fixed-network infrastructure, as far as it is not covered by the current standards work;
- the interfaces, functionalities, and service elements made available by mobile-network operators to independent service providers;
- the interfaces made available to access the intelligent functionalities of the public fixed network(s);
- the interfaces to be offered to mobile networks operated for own use or for use by a closed user group (private mobile networks) to allow interconnection with the public fixed network(s);
- interfaces allowing direct interconnection of mobile networks based on either the same or different technologies.

The Commission intends to initiate a programme of standardisation mandates to the ETSI to accelerate standard development in this area. Those standards should, in general, be voluntary. Only in cases where it is necessary to ensure basic interoperability, should references to standards be made binding.

In its proposed position on a European interconnection policy in the field of mobile communications, the Commission basically regards the existing framework on the European level for interconnection agreements, as developed by the ONP framework directive, the voice-telephony directive and the Treaty competition rules, as sufficient. To facilitate interconnection, however, the establishment of technical standards for interconnection interfaces should be promoted and published.

The ONP framework might have to be adjusted to guarantee mobile operators the right to directly interconnect with other mobile-network operators, both within and between member states. That seems to be the only legislative measure the Commission has in mind to change or to adapt the existing legal framework for interconnection in the field of mobile communications. No further specification of the economic terms and conditions of interconnection agreements seems to be intended.

3.3.6. Evaluation of current European interconnection policy

The fundamental principles for open network provision that have been developed within the legislative ONP framework are a sound basis for a comprehensive European interconnection policy. Although developed for and within an environment of monopolistic provision of voice-telephony and network-infrastructure harmonised conditions, basic principles and essential requirements are also fundamental regulatory conditions in an environment of competitive provision of voice telephony and network infrastructure.

Compared with the regulatory framework and the regulatory rules on interconnection in countries which have introduced competition in the core business of telecommunications, however, the current elements of a European interconnection policy are too broadly defined and don't address all relevant regulatory challenges and problems. They are consistent with quite divergent national approaches in the European Union and, therefore, neither lead to harmonised conditions for interconnection nor facilitate or foster the development of European-wide networks and services.

This evaluation holds in particular where the ONP voice-telephony directive has not received the approval of the European Parliament. Although in our detailed analysis of the ONP voice-telephony directive we conclude that it is not addressing all relevant interconnection scenarios and is not specific enough to provide a sufficient interconnection policy framework, this legislation would have brought the most specific legislative measures for a regulatory interconnection framework in the European Union to start with.

Notes and references

[1] Wissenschaftliches Institut für Kommunikationsdienste/European-American Centre for Policy Analysis: Network Interconnection in the Domain of ONP. Study for DG XIII of the European Commission. Final Report, Bad Honnef November 1994, Annex to Final Report – Country Studies, Bad Honnef November 1994.

[2] This comparison is based on detailed country studies which are included in the study for the European Commission. The country studies on Luxembourg, Spain, Italy, Ireland, Greece and Belgium were worked out by Bernd Ickenroth, the country study on France by Godefroy Dang N'Guyen and the studies on Netherlands and Denmark by Jens Arnbak.

[3] By USF we mean an approach to funding USOs that does not involve access charges and obtains all its financing from other sources.

[4] More precisely, the single-ended charge for interconnection with the local network (terminating or originating) is compared with the tariff for a local call, both measured as a charge per minute.

[5] In the UK, however, an access-deficit contribution (ADC) has recently been imposed on certain calls made by Mercury customers. Including this ADC, the UK figure would be ~100%.

[6] We could not compute the ratio for the fixed/mobile interconnection in Greece as the mode of paying for interconnection there – a percentage of the mobile carriers' revenues – does not lend itself to this kind of comparison.

[7] Commission Directive of 28 June 1990 on competition in the markets for telecommunications services (90/388/EEC), *Official Journal of the European Communities* 16 L192, 24.7.1990, p 10.

[8] Treaty establishing the European Community as amended by the Treaty on European Union.

[9] Council Resolution of 22 July 1993 on the review of the situation in the telecommunications sector and the need for further development in that market (93/C213/01), *Official Journal of the European Communities* 16 C213, 6.8.1993, p 1.

[10] Europe and the Global Information Society, Recommendations to the European Council. Brussels, 26 May 1994.

[11] Council Directive of 28 June 1990 on the establishment of the internal market for telecommunications services through the implementation of open network provision (90/387/EEC), *Official Journal of the European Communities* L192, 24.7.1990, p 1.

[12] Council Directive of 5 June 1992 on the application of open network provision to leased lines (92/44/EEC), *Official Journal of the European Communities* L165, 19.6.1992, p 27.

[13] Council Recommendation of 5 June 1992 on the provision of harmonised integrated services digital network (ISDN) access arrangements and a minimum set of ISDN offerings in accordance with open network provision principles (92/383/EEC), *Official Journal of the European Communities* L200, 18.7.1992, p 10.

[14] Although the Directive was rejected by the European Parliament in July 1994, and although its future is unclear, we regard it as useful to deal with its contents in the same way as before the Parliament's decision.

[15] The directive does not use the term 'access charges'.

[16] Council Resolution of 22 July 1993 on the review of the situation in the telecommunications sector and the need for further development in that market (93/C213/01), *Official Journal of the European Communities* C213, 6.8.1993, p 1.

[17] Council Resolution of 7 February 1994 on universal service principles in the telecommunications sector (94/C48/01), *Official Journal of the European Communities* C48, 16.2.1994, p 1.

[18] Commission of the European Communities, Towards the Personal Communications Environment: Green Paper on a common approach in the field of mobile and personal communications in the European Union, COM (94) 145 final. Commission of the European Communities, Brussels, 27.4.1994.

IV Service Competition or Network Competition?

Telecommunication:
New Dynamics and Driving Forces
A. Bianchi and G. Richeri (Eds.)
IOS Press, 1996

Chapter 7

Privatisation of Hungarian Telecommunications: lessons and questions

Krisztina Heller
Hungarian Telecommunications Company, Budapest, Hungary

Abstract: There is no doubt that the recent breakthrough in Hungarian telecommunications is a big experiment in terms of telecommunications development. People, politicians, investors and analysts are happy and optimistic about the events, although the final proof lies in the future. It seems to be favourable to have a telecommunications industry where everything has been arranged for a long-term development in basic telephone services: capital and service obligations. There are, however, some sceptical opinions about the market potential and the co-existence of regional companies with the new technologies.

1. Introduction

After four years of continuous reforms in the telecommunications sector, the Hungarian Telecommunications Company, which had 80% of the existing telephone market, was privatised on 22 December 1993. The sale was concluded with a purchase price equivalent to US$2436 per access line, the highest price ever paid for a minority stake in a telecommunications company.

Shortly after, on 28 February 1994, the result of the tender was announced where the remaining 20% of the telephone market was awarded to eight companies.

According to the targets laid down in the concession agreements, Hungary, which was ranked the last but one on the European list of telephone density, is soon going to overtake its Central and Eastern European neighbours with a annual increase in access lines of 15.5% up to the year 2000.

On 1 April 1994, in addition to NMT, the existing mobile-service provider, two new GSM operators started their service in the country, reaching 48,000 mobile subscribers, which is by far the largest in the region. On 29 April 1994, two concessions were awarded for providing European radio messaging system-type paging services in the country.

With these promising events in mind, this chapter aims to present an overview of the Hungarian telecommunications story, and makes some suggestions about future developments in that market.

2. The story

2.1. *The late eighties*

It is widely recognised that the communist regime did not support communication between people. Similarly, the planned economy only required the central planning agencies' orders to be communicated to the public instead of allowing market information to flow. As a result, telephone density in socialist countries was low compared with demand and with the general level of the economy, production and human resources.

During the period of soft communism, in the late eighties, the waiting list for a telephone reached as long as twelve years. The first attempts at social reforms in 1989 and the political overturn in 1990 continued to increase the demand for telephones. Demand in terms of the number of access lines and the number on the waiting list increased by half in the course of four years.

2.2. *The first attempts*

This situation was not sustainable and called for radical changes. The first attempts were made with the help of the first World Bank loan in 1988 when US$70 million was borrowed and significant structural reforms were targeted at the Hungarian telecommunications sector:

- In 1989, regulation and operation were split.
- In 1990, telecommunications, broadcasting and postal activities were split into separate companies.
- In 1991, the Hungarian Telecommunications Company was transformed into a joint stock company.

2.3. *The new rate of progress in 1990*

After the political overturn in 1990, various forces promoting sectoral changes gathered new momentum.

Democracy and the market economy, in themselves, demand an increase in the amount of communication and information. As part of the general feeling of freedom, expectations of people in the new regime included getting the long awaited telephones as soon as possible. Last but not least, would-be investors and domestic entrepreneurs, supported by foreign partners, expected to be free to enter this promising industry.

The old monopoly in telecommunications, which had become the symbol of the old and rejected regime, was regarded as obsolete and useless.

2.4. *The government: drafting telecommunications policy*

There was no question that the restructuring process of telecommunications should be based on a new telecommunications policy that was accepted by the majority of the public. Efforts were made to determine the optimal industry structure. Involvement of foreign consultants, financed by foreign-aid programmes, was significant.

The Big Question was the number of service providers and that of monopoly. In the course of three years, various proposals were made for the optimal industry structure. The proposals ranged from a single monopoly, through three regional monopolies and ending with the possibility of as many as 3000 telephone companies, which is the number of the settlements in Hungary. There were proposals for both exclusive and parallel concessions.

The arguments in favour of a multi-player industry structure included:

- rejection of the old monopolies;
- the larger the number of the investors, the more money would be invested;
- municipalities have the right to influence telecommunications development in their regions.

The arguments against a multi-player industry structure included:

- a better price would be paid for the Hungarian Telecommunications Company as a whole; transaction costs being smaller with one sale;
- the negotiating position of the state is better with one strategic investor;
- splitting up the Hungarian Telecommunications Company involves a threat to the integrity and interoperability of the network;

- splitting up the Hungarian Telecommunications Company would involve additional time and cost.

Finally, the Telecommunications Law was approved by the Parliament in November 1992. It left non-telephony telecommunications services liberal, but did not decide the industry structure of telephony; leaving room for deciding it later.

2.5. Hungarian Telecommunications Company: competing with the time

The independent Hungarian Telecommunications Company (MATAV) started in 1990 with a new, young generation of management that was committed to success. Pending the telecommunications policy and legislation, the state did not have substantial regulatory power over the company. Instead, new would-be competitors were founded by would-be investors. By the 1993 their number reached some 50.

In the light of the threatening competition, MATAV tried by any means to get into a better position. These included not only introducing new financial tools, such as bonds and equity warehousing with the European Bank for Reconstruction and Development, but also the founding of joint ventures with other investors, or with its own competitors, for building and financing additional capacities.

As a result, during the three-year plan 1991 to 1993 a digital backbone network was built up, and the number of main lines increased by half.

2.6. The new industry structure

In the middle of 1993, the new, dynamic minister made a big decision. The Hungarian telecommunications network was divided into 54 primary regions. Under the Telecommunications Law, the minister invited the municipalities to indicate whether they wished MATAV to provide services in their region or to initiate a public bid for a new service provider.

As a result, public bids for concessions were invited for 25 primary areas which represented 15% of the telephone market. A 25% Hungarian share in these companies was the minimum requirement. If the public bid for a primary area was unsuccessful then that area would revert to MATAV. MATAV was allowed to bid in these regions.

The rest of the telephone market, including international and long-distance traffic, remained with MATAV and was also subject to public bidding for a strategic investor with a share of 30%. The state retained a minimum share of 51% in this company. The creation of a hub for the telecommunications of the region is also a target for the new MATAV to meet.

Both types of concession promised eight years of monopoly and provided an annual growth target in new lines of 15.5% over the next six years.

Under the umbrella of the concessions with exclusive rights to provide services for subscribers, one can find liberalism in the ownership or operation of the network or both. Some companies own and operate the network for MATAV (for example, First Pest). On the other hand, the local companies are obliged by their concession contracts to buy some services from MATAV (for example, operating support systems).

2.7. The emerging regulatory regime

Preparatory works for privatisation included preparation of the most important pieces of regulation. These included:

- the Network Structure Plan and the Numbering Plan as parts of the basic technical plans;

- procedures for concluding concession agreements;
- tariff regulation and revenue sharing;
- rules of interconnection;
- statute of the regulatory body.

The Network Structure Plan and the Numbering Plan fixed the primary and secondary areas as well as the nodes of the PSTN. This structure plan serves as a basis for the industry structure and for the prohibition of bypass.

The procedures for concluding concession agreements were prepared by following the Council Directive (92/50/EEC) [1]. Emphasis was placed on the professional and financial reliability of those who were bidding, as well as on performance guarantees on their agreements. The process for evaluating the bids is constructed in such a way as to be a transparent.

Tariff regulation has been determined for a longer period of time. Basic elements of the price cap-type regulation include:

- the compound price index (price cap) may follow inflation expressed in terms of the Producers' Price Index (PPI);
- tariff items are also determined for eight years, implementing a rebalancing between local and long-distance tariffs;
- cross-subsidies may be made between regional companies through the Telecommunications Fund (the rules of which have yet to be prepared);
- the revenue-sharing method between primary areas and the long-distance market is fixed for a long time with the underlying principle that two-thirds of the revenue remains within the primary areas as a whole and one-third goes to the long-distance segment.

Rules of interconnection can be found in various pieces of legislation. Technical terms are included in the Basic Technical Plans mentioned above, economic terms in the price decree, and commercial terms in the interconnection decree. Commercial terms were constructed in a way as to protect the users.

Statute of the regulatory body created an inspectorate under the control of the Ministry of Communication. Its basic duty and authority is to control and inspect the implementation of telecommunications regulation, including the licensing and certification of services and equipment, as well as the targets included in the concession agreement. The inspectorate is not directly involved in the commercial affairs of the service providers and users. Although the inspectorate is under the control of the ministry, its decisions can be challenged in the independent courts by its clients.

Applicants for MATAV privatisation were consulted before finalisation of these decrees. Further pieces of telecommunications legislation, such as the terms and conditions of the subscriber's contract, conditions of service licences, and others, will be prepared in the next stage.

2.8. The transaction

The sale process was supported by internationally-known firms of consultants. After intense competition, the consortium comprising Deutsche Telekom and Ameritech was selected with a bid totalling US$875 million. For this, they acquired a 30% share in MATAV. MATAV is going to invest US$4.4 billion over eight years.

Around the half of the privatisation proceeds, in form of concession fees, will go directly into the national budget. Without this amount of money, the deficit of the budget and that of the current balance of payments would have been about 10% and 30% larger, respectively. The investors, however, will get this money back from the state in the form of a tax holiday for the next five years.

2.9. The regional telephone companies

The results of the public bids for the 25 primary areas were announced on 29 April 1994. Forty-one bids had been submitted for 23 regions which were awarded to eight companies. Out of the 25 areas, ten were given to MATAV. The foreign consortia include Austrian, American, French, Israeli, Dutch and Swiss firms.

The concession companies paid US$80 million in concession fees to the government. They are going to invest US$200 million in the development of the telephone network. The consortia are obliged to provide full financial guarantees that the development targets in the concession agreements will be met.

The concession contracts were signed at the end of May 1994. With all the contracts signed, the whole process of telecommunications restructuring could be declared successful.

3. Evaluation of the new arrangement with a view to the future

Telecommunications is an industry where today's investment will have its effect over eight, ten or even 20 years. Therefore, the general trends and their impact on the Hungarian telecommunications market merit some thought, and the main factors call for a closer examination.

These factors are:

- the present market structure;
- the rules created by the state for this market;
- the general trend in the field of technology and services;
- trends in the world economy;
- the playing field within the economy.

3.1. The present structure of the telephone market

In the wake of recent developments, the market in Hungary today is quite complex from the viewpoint of competition.

3.1.1. The legal framework

In legal terms, we now have a horizontal monopolisation and a vertical liberalisation in Hungarian telecommunications.

Horizontal monopolisation means that monopolies for telephone-service provision are regionally distributed over the country. The concession agreements in themselves are tools by which the state transfers its obligation for providing universal service to a private (or private-like) company. The company is responsible for the service targets but has freedom in financing and running the business.

Horizontal distribution of the monopoly was the result of a political compromise with the municipalities and the foreign investors.

Vertical liberalisation means that although the subscribers have their contracts with the monopolistic telephone companies, network operation, network construction, financing, ownership (leasing), and even the non-telephony services have been now fully liberalised.

3.1.2. The market theory

Monopoly in service provision, accompanied by the service obligation is fully justified by the fast development targets (15.5% per year) which would have been impossible under normal business terms. Obviously, this is a justification based on the social-welfare concept as well as on the gap between Hungary and the developed countries.

Vertical liberalisation makes it possible for the players within the telecommunications industry to operate on a viable commercial basis while subscribers are protected through the regulation of these monopolies.

It is worthwhile noting that in terms of market theory, the concession companies are monopolies in their relationships with consumers where unregulated pricing would tend to be related to the price elasticity of the demand. This is the reason for price regulation.

As far as their possible suppliers are concerned, however, they may act as *monopsonies* which means that they have exclusivity as buyers. In this case, producers' prices are the function of the price elasticity of production. Freedom in vertical relationships theoretically gives way to the birth of an optimal market structure in forms of outsourcing, leasing, strategic alliances, etc.

3.1.3. The reality

All of this, however, has been said by certain market theorists. The reality may be different. The *question* is whether a multi-player marketplace will emerge.

Events so far have shown that a very strong tendency for vertical integration can be observed as very few cooperation agreements have been concluded between the companies for leasing, outsourcing and others.

This is a phenomenon which is described by Llerena and Wolff (p 260) [2]: 'When strong uncertainty coexists with a high degree of asset specificity, it is impossible to specify ex ante the whole set of contingent clauses required for contract execution ... This in turn generates enforcement problems ... In this case, it is more efficient to internalise the transaction inside the hierarchical structure of the firm.'

Inter-firm agreements (ie, leasing, outsourcing, etc), however, are undermined in this stage of development by the following factors:

- the large amount of external financing results in a need for power over the whole hierarchy of the business;
- the dynamic environment (technology, economy, policy, demand) imposes a risk which is too large for building long-term commitments.

The lack of legal constraints on the vertical industry structure, however, gives a high degree of flexibility to the players which might be useful at a later stage of development when globalisation of services and technologies will call for 'a network which will be very different from today's networks in both architecture and functionality'. (Egan, p 15) [3].

3.2. The rules created by the state

3.2.1. Dynamism of regulatory philosophy

Hungarian Telecommunications regulation was prepared by implementing European open network provision directives as much as possible. The major characteristics of this regulatory regime are:

- it is *consumer focused* rather than service-provider focused, because the scope of the Telecommunications Act and related regulations are the technical, financial and commercial terms and conditions of providing the services;
- there is *no asymmetry* in regulation because it is based on objective criteria that are uniform for those providing the given services;
- there are as *few sector-specific rules* as possible and sector-specific regulations are applications of the pieces of general legislation.

As it was stated earlier, by the time of the transactions the major pieces of regulation required for long-term commitment by the investors and for their interconnection had been prepared. Would-be investors were consulted before finalising the decrees.

However, while all of this was enough for successful transactions, it was not enough for the day-to-day functioning of the industry. There are a lot of basic *questions* to be answered by further pieces of legislation as soon as possible, such as:

- What are the rights of the subscribers and how they can enforce them?
- What are the rules of network access?
- How will standardisation work?
- What are the conditions for getting licences for services, networks and equipment under the new circumstances?
- What are the rules for leased lines?

In the regulatory vacuum, life is going on: players in the telecommunications arena have started to learn how to cooperate. The *question* is whether the regulatory vacuum will lead to a quasi-independent regulatory regime or to disorder.

3.2.2. Price regulation

The thumb-rules of price regulation presented earlier represent a major step in the direction of a good price-cap regulation, but it can hardly be thought that they will work without problems in a changing and complex environment. There are a number of *question marks* over price regulation.

The *first* question mark is over PPI. Countries using a price-cap regulatory regime implement the RPI (retail-price index)-minus-X model, where X stands for productivity improvement.

The use of the PPI is justified by the fact that given the change in macroeconomic structures, it is more stable than the RPI. In addition, PPI in Hungary has been lower in value than RPI, which is keeping national inflation lower. The Hungarian regime, however, does not have a productivity item in the price cap. This is due to the fact that under the dynamic circumstances like this, it is impossible to make reasonable estimates of a reasonable productivity.

The question is: does PPI reflect inflation and the productivity improvement at the appropriate level in the long run? Will it produce a reasonable return on the money invested?

The *second* question mark is over the rebalancing between the long-distance and the local market. For the time being, nobody is familiar with the costs related to these markets. Furthermore, the development trends of the international telecommunications markets will be also of significance, and these are also unknown.

The *third* question mark is over the cross-subsidisation between the regional companies through the Telecommunications Fund. Rules for the functioning of this Fund have not been specified so far. Personally, I guess that they might be related to the use of other regional-development funds. The big dilemma is whether the distribution of the funds available is made by the regulator (centrally) or by the players themselves (by some sort of agreement). The answer to this question seems to be a difficult political process.

On the whole, the highly-segmented market implies a lot of cross-subsidisation which is not internalised but externalised between and among the organisations and markets. This is not only hindering the functioning of the marketplace, but also may hide the effects of an inefficient way of doing business.

The *fourth* question mark is over the revenue-sharing method of 2:1 in favour of the local traffic, which seems to be too mechanical for a functioning marketplace, and gives little incentive for investing money for a higher return.

3.2.3. Need for incentive regulation

The concept of incentive regulation is usually used for characterising *price-cap* regulation as opposed to that of *rate-of-return*. However, there are authors who use this concept in a broader sense for a regulation that is restricted to areas where competition cannot or does not work [4].

There is a fairly common understanding among the Western authors that developing countries in Central and Eastern Europe need a non-incentive regulatory method, for example, rate-of-return price regulation, instead of that of price cap. Their main argument is the lack of satisfactory certainty in the investment.

In the Eastern environment, however, this argument works in the other way round: incentive regulation under the circumstances of insufficient certainty can serve as a major driving force in developing the market and the overall economic activity. Furthermore, incentive regulation, in fact, is privatising not only the ownership but also the risk of doing business. This is really a divestiture and privatisation of the risks.

The risk to the investors, however, can be under their control. Incentive regulation implemented *ex ante* the competition leads to the rules working as targets for the telecommunications business rather than as corrections for market failures. In this case, the question for the would-be investor is whether the targets are reasonable and justified and, later, whether there is room for improving the conditions.

Under the legal structure of Hungarian industry, there might be enough flexibility for finding economic solutions. Regulatory solutions are included in the concession agreements.

3.2.4. Privatisation or trans-nationalisation?

In a situation where investors are responsible for the service obligations and have to bear a lot of risks the question arises: why is it favourable for them?

Answering this question would lead us far away from the purpose of this chapter. It is appropriate, however, to mention that 'private investors' have not totally been private, but rather state-owned, for example, Deutsche Telekom. Looking at the foreign investors getting involved in Hungarian telecommunications, one finds a substantial involvement of a number of states.

3.3. The general trend in the field of technology and services

The latest trends in the telecommunications market are now quite clearly visible. They may become significant for the future of Hungarian telecommunications as well.

3.3.1. The trend of globalisation

The broadband information superhighway is playing an ever-increasing role in communications and in society. The United States, as well as the European Union and Japan are creating a global information superhighway or infrastructure.

The new trends are closely related to the fact that digital technology will soon be dominating all telecommunications services. The services are diversifying in terms of:

- type of information (image, voice, data, etc);
- degree of mobility (cordless devices, mobile telephone, aircraft telephone);
- type of communication (call, conference, etc).

The growing number of services on offer goes hand-in-hand with a growing number of applications, and the technologies that have been implemented are also diversifying. Finally, integration of technologies also can be observed, for example, fixed with mobile, voice with others.

Egan summarises this trend in a compound way: 'The networks of the future exhibit different structure (architecture) and functionality. These networks are integrated and technologically sophisticated, exhibiting broad functionality capable of simultaneously providing many distinct customer services via the same access line.' (Egan, p 17) [3].

With various services using various types of networks, a new, matrix-type industry structure is emerging which Miller calls a 'virtual marketplace' [5].

The *question* is whether the actual industry structure and regulatory framework is compatible with the trends of the future.

3.3.2. *Compatibility with the trends of the future*

From the viewpoint of this new type of industry structure, the Hungarian arrangement is promising. The flexible vertical structure will easily be able to develop into a matrix-structure. Recent experiences, however have not so far exhibited any movement towards the network of the future.

The recent experiences with the new regional concession agreements show that new entrants are reluctant to outsource some of their activities, such as operations. This behaviour perhaps reflects the availability of economies of scale and scope between operators and service provision.

The structure of Hungarian telecommunications, with a number of regional companies, is not favourable for the development of an integrated network. Some of the regional companies are too small to be able to implement the latest technology. Therefore, the rapid expansion of a regional network might leave an engineering and financial legacy that could inhibit the development of the information superhighway.

Clearly, overlay networks will have to be constructed which will break the present structure of the regional companies.

The *question* is: from where will this overlay network originate? There are three options:

- the regional companies;
- the Hungarian Telecommunications Company with the long-distance network;
- a cross-border network.

The outcome will be decided by the capital and market strengths and potentials but preparation for this battle, which will take place in eight years (after the expiration of the monopolies), will start in the near future.

3.4. *Trends in the world economy*

Hungarian telecommunications policy, investments and financial sources are closely related to the country's political and economic relationships with other countries.

3.4.1. *Global trends*

Within the next ten years the most important factors in these fields will surely be the following:

- the obligations arising from the Association Agreement with the European Union;
- the completion of the Uruguay round of GATT and the obligations arising from that contract;
- the leading role Hungary would like to play in Central and Eastern Europe's economy and telecommunications.

The telecommunications policy that is adopted by the European Union will have a decisive role. Although full liberalisation of basic telephone services is not expected before 1998–2003, major changes towards deregulation and commercialisation of the market have already been implemented.

One of the major targets today is the liberalisation of cross-border services within the EU. Several measures have been taken with a view to creating broadband pan-European networks and services, and international traffic of the closed-users' group is also going to be liberalised. The latest technologies (mobile and universal personal communication) are an increasingly important factor in the progress towards liberalisation.

The European Union, however, has not prepared a comprehensive policy for integration with the Central European countries.

3.4.2. The challenge for Hungary

Under the concession agreement, MATAV has a monopoly on international traffic for eight years. It seems, however, that this monopoly will slowly but surely be eroded over the medium term. GATT will accelerate this process.

Erosion is necessary if we are to assume a decisive role in Central and Eastern Europe because, having reciprocity in view, it is impossible to open the market to the East while being closed to the West.

What follows is that if Hungary wants to survive against the increasing international competition, the necessary steps, including implementing new technologies, have to be taken very soon (for example, Spain has decided voluntarily to liberalise voice telephony in 1998).

The *question* is: how long does it take to arrive at a need for international liberalisation, and in what form it will happen?

As Pietras points out, connecting to Europe via an 'interface', or, on the other hand, to be integrated into it, implies a major difference in terms of policy, technology, business, etc [6].

One can hope that having invested in Hungarian telecommunications, companies will soon be 'mentally nationalised' and will help Hungarian telecommunications and society to become a 'global franchise' to Europe [6].

3.5. The playing field within the economy

The telecommunications industry is in need of an economic environment that could serve as a proper basis for the successful development of the network infrastructure and the industry itself.

The playing field for Hungarian telecommunications will be determined by the following factors:

- the cost of leaving the marketplace;
- the profitability of the industry as a whole;
- the economic viability of the individual organisations providing services.

If the telecommunications industry as a whole is not profitable, there is no chance for successful cross-subsidisation and profit-sharing between organisations. But if the industry is profitable on the whole, then the participants will be given a sufficient playing field and a sound market may emerge.

3.5.1. The cost of leaving the market

Theories on contestable markets are frequently used for justification of free (liberalised) entry to the market as a means of liberalising the telecommunications industry. These references hardly mention, however, that rules of contestable markets work only if and when the cost of leaving the marketplace is around zero. The truth of this statement became evident in Hungary during the past three months.

According to the Hungarian Telecommunications Law, concession agreements concluded between the state and the service provider, and new entrants and the incumbent service provider have to agree on the economic and financial terms for transferring the existing parts of network from the old company to the new one.

Practice has shown that this approach is hardly viable. The new players do not need the old assets, which, in turn, represent value for the incumbent. The standpoints of the relevant parties are too far apart to be expected to meet.

This fact highlights the problem of managing the structural changes taking place in the telecommunications industry: the shift from analogue to digital, from fixed to mobile, and, in the future, from single to multimedia.

The *question* arises: if new entrants represent the new technology, who will bear the cost of the old technology leaving the marketplace?

This seems to be a general structural-change phenomenon experienced at various times in history (for example, metallurgy, energy, etc) which, in this case, calls for management by a telecommunications policy.

3.5.2. Short-term prospects

In the first two or three years, general profitability is not likely to deviate significantly from the anticipated figures. However, a few organisations may find themselves in awkward financial situations as a result of having misjudged their own resources, the market potential or the economic environment.

In such cases, the government must intervene as a crisis manager using the instrument offered by the Telecommunications Fund, but in order to support the consumers and not the failures of the service provider.

In other words, such subsidies should be available only if absolutely necessary for keeping the consumers supplied, and only for a short period of time.

3.5.3. Medium-term prospects

The industry's full market potential will be more important in the medium run. This raises the following questions:

- What will the impact of the recession be on demand and the subscribers' ability to pay?
- Is the market potential large enough for repayment of the concession fees paid into the national budget at the time of the transaction?
- Will tariff regulation prove effective in controlling the rate of increase equal to the PPI?
- How will the impact of devaluations of national currency be balanced?
- How fast will erosion of the international monopoly be?

It is an easily-conceivable scenario that the general profitability of the industry will be relatively low. In this case, the regulator will have to reconsider any of the tariff ceilings, concession fees or taxes.

In the medium term, revenue-sharing between companies must be settled within co-operation mechanisms, instead of the 2:1 principle, the common interest being an increase in the throughput of the overall network.

In the case where some of the telephone companies are unable to survive, the market mechanisms (bankruptcy, mergers and aquisitions) would maintain the dynamics of the market. In such cases, the Telecommunications Fund may be used as an intervention fund for maintaining universal service.

4. Long-term speculations

As far as long-term prospects are concerned, one cannot forget that all over the world, telecommunications companies are in the process of horizontal integration (for example, in Finland). At the same time, Hungary is splitting its market horizontally.

It is hardly conceivable that this horizontal industry structure will remain in the long-term. Mergers can be expected which might be 'supported' by the bankruptcy of some companies. The first set of mergers, in form of that of the owners in holding companies, could be observed at the time of signing the concession agreements.

Besides, very soon – the sooner the better – we shall have to face the question of how to switch over to the next, higher level of technology, into the world of the information superhighway. This is an issue that will be important for the role Hungary plays in the world market. Moreover, the role played by Hungary in the world economy will determine the nature of information-superhighway developments in Hungary.

In the long term, the key issues will be:

- how the users of the information society prosper;
- when the information superhighway will be available.

Under the macroeconomic influences referred to above, it is not unlikely that by the year 2000 the Hungarian telecommunications market will have bled out and be left undercapitalised. This would imply that new sources and schemes of financing should be found.

One must not forget that, due to the shift of power from the hardware to the software side, the telecommunications industry will lose power and will have less influence over the outcome of the situation. Under the new circumstances, it is possible that users of telecommunications and information services will have more power over the whole telecommunications policy and will get some sort of support from the government, possibly in the forms of subsidies, credit facilities and other regulatory means.

My vision is that in the best case, Hungarian society and the economy will be an integral part of Europe where telecommunications will be developed in the global European framework.

In a less favourable scenario, Hungarian telecommunications will be 'plugged' into the European telecommunications world via financial and technological 'interfaces' – re-launching the whole story of the reforms of the 90s.

5. Conclusion

There is no doubt that the recent breakthrough in Hungarian telecommunications is a big experiment in terms of how the telecommunications industry in Central and Eastern Europe is able to catch up with the requirements of social and economic development. People, politicians, investors and analysts are happy and optimistic about the events, while final proof lies in the future.

From the telecommunications policy's point of view, it seems to be favourable to have an industry where everything has been arranged for a long-term development of basic telephone services: capital, service obligations, etc. This would allow the policy-makers and regulators to keep their hands on the value-added services and new technologies. This is, however, the optimist's point of view.

There are, however, some sceptics asking:

- Will the economic recession maintain the growth in demand?
- Is there enough potential in the market to repay the large concession fee paid into the national budget?
- Will foreign investors find the regional companies' market big enough to be financially viable?
- Will new developments in technology be appropriate for co-existence with the old structure of the telephone companies?

Clearly the answers lie in the future.

Notes and references

[1] Council Directive 92/50/EEC, Relating to the coordination of procedures for the award of public service contracts, *Official Journal of the European Communities* No L209, 18.06.1992, p 1.

[2] P. Llerena, & S. Wolff, Inter-firm agreements in telecommunications: elements of an analytical framework. In G. Pogorel (Ed.), Global Telecommunications Strategies and Technological Changes. North Holland, Amsterdam, 1994.

[3] B. L. Egan, Information superhighways: the economics of advanced public communications networks. Artech House, Boston, 1991.

[4] J. A. Ordover & R. W. Pittman, Competition Policy for Natural Monopolies in a Developing Market Economy. In OECD/World Bank Conference on Competition and Regulation in Network Infrastructure Industries, Budapest 28 June–1 July 1994.

[5] D. Miller, Cooperation in a virtual marketplace. In The Networked Economy, Proceedings of the Conference held in Paris, March 1993. CMP International Conferences in association with *Communications Week*, 1993.

[6] J. Pietras, Awakening to Services and Networks: three scenarios for Eastern Europe. In Networld Order Scenarios, *Project PROMETHÉ Pespectives* No.17, Paris, June 1991.

Telecommunication:
New Dynamics and Driving Forces
A. Bianchi and G. Richeri (Eds.)
IOS Press, 1996

Chapter 8

Fostering and using the highways and byways of Europe

Richard A. Cawley
European Commission, Brussels, Belgium

Abstract: This paper deals with the general logic and the practical aspects of the preparation for full telecommunications liberalisation in the European Union in 1998. The political agreement on full liberalisation enables a coherent legislative package to be prepared, modifying the 1990 services directive to encompass infrastructure, public voice telephony and indeed the other areas which were initially excluded – satellite, mobile and cable television. At the same time a number of very practical questions arise, including the framework for licensing, interconnection, universal service and other key areas such as numbering. In the longer term, the other major exclusion from the services directive will also have to be considered, that of broadcasting.

1. Introduction

The theme of this session of CPR '94 is 'Preparing for 1998 (telephone competition) – service competition or network competition?'. The simple direct answer to this question is an unambiguous 'full network competition' from 1998 but there are some additional points to make.

The first is that nothing prevents individual countries liberalising earlier. This is already the case for the UK, Finland and Sweden and moves are afoot in the Netherlands and Denmark.

The second point is that prior to 1998 there will be significant telephone competition, which is infrastructure-based, of a different form, ie, mobile cellular. This is both a conjecture, based on the fast developing market situation in GSM and DCS 1800, and one which is based on the proposed Article 90 directive on mobile and personal communications [1].

The third point is that the Commission has proposed a faster liberalisation of infrastructure for liberalised services and has proposed a framework within which services liberalisation and infrastructure liberalisation can be linked. This is the background to the proposed liberalisation of alternative infrastructures in advance of 1998 via the Article 90 directive on cable television [2] and via the proposed Article 90 directive on full competition [3].

This chapter examines the logic of the policy on infrastructure liberalisation including the timing of liberalisation and the accompanying regulatory measures. It argues that the Commission has now proposed a coherent framework of measures for full liberalisation. It also argues that full and effective facilities-based competition is required in order that the required pricing reform in the sector can take place. Without this reform and the necessary stimulation of telecommunications usage through lower prices at the margin, there can be no real take-off of the information society.

Tariffs, pricing and new services can be seen as the indicator of how far the telecommunications sector has moved towards a new commercial approach. In the broader European and international context and on the basis of both political agreements and legal measures that have taken place in the European Union, substantial progress appears to have been made. The current reality in terms of pricing and service quality is at best patchy in many parts of Europe. Recent reports or surveys by the OECD, INTUG/ECTUA (international telecommunications user group/European telecommunications user association) and indeed ourselves at the European Commission show significant disparities of performance across Europe and between Europe and North America.

The OECD work (the 1995 Communications Outlook) [4] is a good example. It clearly shows that a performance gap has opened up between those countries which have liberalised faster and those which have not. It is not simply that the first group has rebalanced tariffs faster; tariffs have actually fallen more in real terms.

More seriously, the tariff innovation and reform that is required to stimulate usage and new communications applications is only happening at the margins – in mobile communications, in some indirect competition on fixed networks and on the Internet where cheap or flat-rate tariffs are stimulating massive use of data communications.

At the moment, there are no real incentives for many of the traditional operators to undergo pricing reform and encourage usage of existing and new applications. Full competition will change all that and is a necessary condition for fostering and using the highways and byways of Europe.

2. General market and pricing background

Recent work by the OECD (the 1995 Communications Outlook) and surveys by INTUG/ECTUA and the European Commission show a disparate situation for tariffs and pricing. In general, rebalancing is taking place and tariffs are declining overall. However, the speed with which this is occurring and the amount by which tariffs are falling overall varies considerably. The situation can rightly be described as two or three speed Europe, normal, slow and hardly moving at all.

At the international and global level, the slow response of excessive international tariffs to lower costs and greater usage has led to all kinds of attempts to beat the system and introduce competition by the back-door. Examples are call-back, various card-based services, private networks, 'break-in and break-out', and the latest is software which allows two-way voice over the Internet. For all the talk of information haves and have-nots at the 1995 G7 conference, the one thing that might tempt my aunt onto the Internet is the thought of being able to chat with relatives in the US and Australia for the price of a local call.

The general approach of the Commission has been to argue for reform but recognise that competition is the only long-term solution. You must remember that the Commission can not legislate on prices – only the agriculture sector does that.

Having said that, the European Union has adopted legislation proposed by the Commission on the pricing of leased circuits. This is because we are in a period of semi-competition. Many services are liberalised but that is generally not the case for infrastructure and widespread effective competition in underlying infrastructure has yet to emerge. New market entrants, therefore, need a legal safety-net to ensure that they can get reasonable access to network building blocks in order to run their business. In Europe, the UK is furthest along this transition-to-competition path but even there things are not fully contestable. That is why controls on British Telecommunications are still in place, both on its retail tariffs and on its interconnection and its leased-circuit tariffs.

The essential responsibility for enforcing European legislation on matters such as leased circuits, as is also the case for interconnection and for licensing, lies with independent national regulatory authorities (NRAs). That obviously implies that there has to be a NRA to begin with. But there are recourse procedures or alternative avenues. National courts are one.

The competition authorities in Brussels are another. For the specific case of leased lines, the open network provisions (ONP) and ONP Committee can also seek a settlement, as in a recent case concerning the supply of a leased circuit to a reseller in Spain. And a final sanction is the European Court in Luxembourg, although this can be a lengthy process.

The general principles in the legislation for the tariffing of leased circuits (and this will also be the case for interconnection), are fairness, transparency, non-discrimination, pricing based on objective criteria, (for example, linked to clear, transparent and appropriate accounting methods) and cost-orientation.

The cost-orientation policy of the Commission has been much maligned and mis-represented, so it is worth taking a moment to try and disperse the myths. The idea behind cost-orientation is to mimic competition or promote fair competition and thereby promote efficiency. Competition tends to drive prices towards underlying costs. Hence its appearance in a legal instrument like the leased-lines directive. Because of the regulatory situation in much of the European Union, competing service operators are obliged to go to the incumbent telecommunications operators (TOs) for the leased lines on which they run their services. The pricing of leased lines should therefore be non-discriminatory in nature and be based on transparent and objective criteria that can be linked back to costs.

Conversely, cost-orientation need not be the over-arching factor in market areas subject to competition. There, operators or service providers should be free to experiment with pricing strategies, within the confines of general competition policy. This is particularly the case for new types of products. That is the situation for other economic sectors and should also be the case for telecommunications.

What about cost-orientation and the retail telephone service? Well it is not intended that pricing mirrors element-by-element cost structures in the network. One example is the balance between the price paid for access to the network and usage prices. The Commission has firmly encouraged both the rebalancing of tariffs and the reduction of tariffs towards more efficient levels. It has done this because it recognises that competition will take some time to emerge, that cost structures in telecommunications are changing and that rebalancing within certain limits provides benefits to many customers and assists efficient competitive entry.

Having said that, a number of combinations of access and usage charges are consistent with competition and choice. This is because there is some leeway between average incremental cost and stand-alone cost, effective lower and upper bounds on an operator supplying a mix of services. A commercial operator would be unlikely to align its access charges strictly to fully-allocated access costs in much the same way as a supermarket does not charge people to enter the store. What a commercially-minded operator is likely to do is to develop a mix of pricing structures to suit different types of customers (although this should be in a non-discriminatory fashion), in much the same way as utility companies (volume reductions, low-access/high-usage tariffs for small users), airlines (pricing linked to demand elasticities) and railways (season tickets) are already doing. This is why the Commission called for cost orientation *and* flexibility in dealing with retail telecommunications tariffs; cost orientation so that prices decreased and better reflected the cost structures of modern networks, and flexibility in recognition that we are moving towards more optional pricing and more discriminatory (in the good sense) pricing.

At the same time, I believe that you are not going to see the full benefits of innovative pricing until effective competition emerges. For the moment the mobile operators are taking the lead in this area. The traditional operators will simply tell you that they are not allowed to do it when the simple truth is that for the moment, those who are responsible for pricing strategy often do not have a big enough say in the company and in any case the billing systems are not always up it.

At the moment, leased circuits are the only real example of flat-rate tariffing in Europe. The small degree and largely inefficient flat-rate pricing of local telephony in Europe has largely disappeared. But that does not mean that flat-rate pricing as an option will not make a re-appearance in the area of interconnection and in retail telephony generally.

In any case, for the moment while waiting for effective competition to emerge, it is just as important to focus on the general level of tariffs. That is also why the Commission has argued the case for price-caps plus some pricing flexibility in this transition phase to competition. Even if some people are lucky enough to be paying cost-oriented tariffs, in most countries they are not yet paying *efficient* cost-oriented tariffs. But it is important to remember that the Commission has no power to impose price-caps. And its power to impose cost-oriented tariffs for leased circuits has not yet been successfully tested, even if some competitive mobile operators have been able to successfully appeal to their national regulators based partly on this piece of legislation.

The last general point is made with an eye to the future. Many of today's pricing structures are a legacy of the past and have very little to do with the underlying cost structures of modern telecommunications networks. To see that, you only have to compare tariff structures in the UK with those in Greece or to examine the direction in which the tariff offerings of PCS operators are moving.

Beyond this, continuing technological and service development will have a profound impact on pricing in the longer term. The potential convergence of telecommunications and broadcasting networks as exemplified by the gradual emergence of some multimedia services, will lead in the longer term to a shake-up in tariffing. At an abstract level, the question will be what bandwidth tariff reductions will operators offer to migrate customers, initially business ones, to new services. At a more concrete residential level, it will come down to the relative tariffing of telephone and, say, video-on-demand. But a major question will be how much are people (both business and residential users) willing to pay for content. Residential or non-business customers seem happy to pay for entertainment but the case for many other multimedia services still seems to highly uncertain. What is clear is that transport costs or at least marginal or transport-usage prices will continue to decrease for all bandwidths and this is bound to encourage new types of service usage.

3. The political agreement on infrastructure liberalisation

In many ways, the Commission was relieved to get the political decision on full infrastructure liberalisation by 1998 in November of last year (1994), even if the intermediate step proposed by the Commission of allowing free choice on infrastructure for services already liberalised (the so-called alternative infrastructure proposal) was not agreed as it was only supported by six out of the 15 member states.

The notion of planning for liberalisation in 1998 based on services only was rather an artificial one. In addition, it is very difficult to construct framework legislation when services or infrastructure are partially liberalised. This has already proved to be the case with questions of what constitutes value-added services (liberalised from 1990) or what constitutes 'everything except public voice telephony' or 'closed-user groups' (liberalised from 1993). Similar difficulties have arisen in the satellite area where, in addition to making this service distinction, one has to distinguish between services and infrastructure.

The November 1994 agreement now allows the Commission to plan an overall package of reforming measures to deal with full liberalisation.

Before I go on to the general logic of the reform process and what it means in terms of modifying Union legislation or adopting new legislation, (and what that then implies for the companies who will be operating in the sector), it is worth recalling to what extent competition on voice telephony based on competing infrastructures, ie, mobile or radio technologies, is already occurring.

4. The increasing impact of wireless technologies on the regulatory framework

In 1994 the Commission published its Green Paper on mobile and personal communications [5]. This resulted in substantial support (including from a significant number of incumbent fixed operators) for the proposal that mobile operators should have a free choice with respect to the installation of their networks. This includes free choice (subject to rights of way, planning and environmental considerations, etc) on self-provision, use of third-party infrastructure and sharing of sites where appropriate, including free choice on transmission media.

The comments received on both the mobile Green Paper and the infrastructure Green Paper [6] [7] demonstrate that it makes no commercial sense to artificially handicap competing mobile operators and that in any case, the burden of regulation, where artificial infrastructure or interconnection constraints are imposed on competing operators, will be excessive. This has already proven to be the case in France and Germany where regulators have been obliged to intercede and modify the initial constraints (in the form of obligations to lease lines at public tariffs) on the second cellular operator.

There were a number of policy proposals in the mobile Green Paper but the main ones that are relevant in the context of developing telephone competition and whether it is service or infrastructure-based are the following:

- Competitive provision of GSM or digital service based on free choice of infrastructure.
- Award of at least one DCS 1800 licence per country.
- Mobile operators to be allowed to interconnect their networks directly across Union borders.
- Freedom from 1998 for mobile operators to apply for licences, to provide services outside their licensed technological and geographical base.

The third proposal above would lead to significant competitive pressure on long-distance and cross-border telephone services in the European Economic Area, and for that reason has a logical link with the 1998 deadline.

The fourth proposal would herald a genuine development in personal communications to the extent that operators could offer services across Europe, based on combinations of underlying fixed and wireless infrastructures.

4.1. Well over ten per cent of subscriber lines will be wireless-based in 1998

In order to emphasise the growing importance of wireless-based markets, this section reviews the current and forecast subscriber-base for cellular users. At the end of 1993, cellular subscribers in Community and EFTA countries numbered 8.8 million compared with 174 million fixed lines. The total for the former subsequently advanced to over 14 million at the end of 1994. The overall ratio of cellular to fixed subscribers is now in excess of five per cent. The awarding of competing GSM licences in Italy, Spain, the Netherlands, Ireland and Belgium is likely to reinforce the growth of cellular subscribers, particularly if prices continue to fall or attractive tariff packages for lower-usage subscribers or both appear.

It is possible to assemble growth forecasts for fixed and wireless lines for the end of 1997, bearing in mind current growth rates, planned investments and penetration rates for more mature cellular markets. The forecast for fixed lines is to some extent an increasingly arbitrary one, bearing in mind the spread of integrated-services digital network lines and dedicated connections for larger customers. However, it seems likely that about twelve per cent of lines (30 million relative to nearly 200 million) will be wireless-based subscribers in the Union and EFTA countries at the beginning of 1998.

A majority of these countries have already decided to allow competing operators a free choice on the establishment of infrastructure, subject to rights of way, planning and environmental rules. This implies that, irrespective of the more recent agreement on the opening of infrastructure to competition for fixed telephony, significant infrastructure-based competition will exist in 1998. Moreover, because of the number of mobile operators that are now licensed, it is important that an interconnection regime exists which satisfies the requirements of mobile as well as fixed operators in Europe.

5. Preparing for 1998 – a framework for both services and infrastructure competition

Following the publication of Part II of the infrastructure Green Paper, the Commission undertook consultations with the sector on the major issues raised, in particular the framework for interconnection, licensing and universal service and published its report on the consultations in May 1995 [8]. The Commission would in any case have had to deal with these issues in the context of the 1993 agreement to liberalise services in 1998.

The advantage of the November agreement to liberalise infrastructure is that the issues can now be dealt with in the overall context of full liberalisation. There is actually very little time in legal terms to prepare for the 1998 liberalisation. The Commission must come forward with its legal-reform package by the end of 1995 in order to give enough time for the European Council and the European Parliament to adopt the legislation within the deadlines and for this to be implemented by member states. In fact, for some countries with slow legislative processes, they almost need to start the process of national legal reform now in order that full competition can begin in 1998.

The one uncertain issue in all of this, is whether the Commission fully succeeds in coming forward with an intermediate infrastructure liberalisation step before 1998. Six of the fifteen member states have lobbied for this. Some have argued that effective competition cannot start in 1998 unless other infrastructures are already up and running. In addition, mobile operators have lobbied for earlier opening for mobile networks and some cable-television operators and some utility companies are also in favour of an earlier move. The arguments for earlier opening are very sound. The problem is that a majority of member states and a majority of incumbent operators are against it and it has become increasingly difficult for the Commission to come forward with binding Article 90 legislation without full political support, including that of the European Parliament.

I will come back to the legal and practical logic of liberalising services and infrastructure at a European level in a moment, but first I want to examine the three key issues in the 1998 preparations: licensing, interconnection and universal service.

I think it is worth looking at these because although they are not directly related to pricing and market development, they are crucial issues for operators who may be extending operations to other European countries, or facing additional competition in their own country, or may be moving into the telecommunications business for the first time, or who as regulators may be dealing with such matters in the future.

5.1. Setting the right regulatory framework

The aim of liberalisation in the telecommunications sector is not to make way for a business 'free for all'. Some basic rules of the game are necessary to ensure that the market can operate in a way that benefits everyone.

That is why the Commission published Part II of its infrastructure Green Paper and consulted the sector in particular on the three major issues.

5.2. *The licensing framework*

The general principle must be that any company should be able to provide any service to any customer. However, there will be cases where member states, who will retain their responsibility for issuing licences or authorisations in the telecommunications sector, will be justified in limiting the number of companies operating by restricting the issuing of licences.

These circumstances, the Commission believes, should be interpreted narrowly so that licences can only be refused or limited in numbers because of access to scarce resources, such as spectrum. We believe that it is unreasonable that licences should be refused for environmental reasons. Rather, conditions can be attached to licences in order that planning and environmental concerns are safeguarded.

In the comments on the Green Paper, the Commission has found support for a framework on licensing which keeps bureaucracy to the minimum but which nevertheless leads to transparent licensing procedures in member states with effective recourse mechanisms.

The Commission intends to develop licensing guidelines which strike a balance between the rights of licence holders and their obligations and is effectively a tiered approach. This means that licensing procedures and the conditions attached to licences must reflect the extent of the rights and obligations which the licence will convey. This is likely to result in different classes and types of licences within national licensing regimes.

The consensus that has emerged from the Green Paper consultations is that the sole reason for limiting the grant of licences for the provision of telecommunications infrastructures and services should be the existence of physical limitations, imposed by lack of frequency spectrum. This is one of the essential requirements recognised in Union law. Other essential requirements found in the Union's regulatory framework apply to services provided over networks rather than to the infrastructure per se.

With regard to the possible extension of the essential requirements to include restrictions based on environmental policy (such as rights of way), the Commission accepts that rights of way do not constitute a physical limitation on the granting of infrastructure licences and can not be used to justify a priori restrictions on licence numbers.

With respect to the number of service providers, no regulatory limitations should be imposed and they should not be subject to individual licensing.

Although no complete consensus has emerged on the issue of public-service obligations, the Commission supports the clear trend in comments received that the issues which 'trade regulations' seek to address are issues that can be tackled within the terms of a licence or during selection procedures, rather than being a reason for justifying an a priori limitation of the number of licences granted.

In order to establish the grounds which may be used to limit the number of licenses, as well as the categories of conditions which can be attached to licences, for the provision of public and private telecommunications services, the Commission has proposed amendments to Directive 90/388/EEC (the services directive) [9] and will also propose a directive on licensing.

5.3. *Interconnection and inter-operability*

The second aspect is interconnection. A framework is needed to deal with interconnection between existing networks and new ones, between fixed and mobile networks, and which covers all the possibilities that may occur at a European level. This means that a common set of rules and conciliation mechanisms is needed to resolve conflicts when things go wrong.

The European Court has endorsed the application of the Union's competition rules to the telecommunications sector and competition rules do apply to interconnection agreements.

However, many respondents to the Green Paper have supported the need for a clear interconnection framework which complements the application of competition rules.

It will be via an overhaul of the ONP framework that the ground rules for interconnection in a more competitive environment will be established. The current ONP framework is already creaking under the strain of developments, in particular in the member state which has advanced most quickly on competitive provision, for example, the application of the ONP leased-lines directive [10] to a limited group of operators in the UK was contested in 1994 by British Telecommunications.

However, ONP is not something to be thrown away in the period leading up to 1998. It is still required in the current environment where service providers, mobile operators and consumers are dependent on ONP principles to obtain reasonable access to and use of incumbent networks. For example, in July 1994, the Conciliation Procedure of the ONP Committee was invoked to settle a complaint about access to leased circuits in Spain.

Currently, the ONP framework applies to telecommunications organisations with special and exclusive rights for providing certain public telecommunications networks or services. After 1998, these rights will be lifted and many more organisations and companies will enter the European market in one way or another. A flexible approach will be required but the Commission is proposing binding legislation in the areas of voice telephony, leased lines and interconnection.

The Commission attaches considerable weight to the concerns of industry, users, potential competitors and service providers for the need for uniform application of any interconnection framework; for its effective enforcement by national regulatory authorities and for the need for the regime to be in place by 1 January 1998. For these reasons, the Commission has adopted a proposal for a directive on interconnection [11] in the context of ONP, which addresses the areas of:

- *Rights and obligations of service providers and network operators with regard to interconnection, in accordance with the principle of proportionality.* The environment envisaged is one where any organisation has the right to request interconnection, but where some organisations have stronger obligations than others to accept interconnection requests. These stronger requirements will be linked, in particular, to the need to ensure European-wide offerings for leased lines and voice telephony.
- *Principles for interconnection charges and accounting systems, including universal-service contributions.*
- *Transparency of agreements.* Interconnection agreements should be made available to the national regulatory authority and, with the exception of those clauses containing confidential technical or commercial information, should normally be published.
- *Role of the national regulatory authority.* At a national level, member states will be required to set up independent procedures for dispute resolution. At the European level, a dispute resolution procedure will be proposed, based on the conciliation procedure already in operation under the ONP leased-lines Directive 92/44/EEC [10].

5.4. Universal service

Thirdly, but probably most important in political terms will be the framework for financing universal service in a competitive environment. The first point is that competition and customer-focused operators *can improve universal service*.

Competition and public service are compatible. In fact they may be mutually reinforcing, as long as it is recognised that public service does not necessarily mean either monopoly provision or provision by a public operator.

There are three important elements to universal service: scope, costing and financing. Universal service clearly has a key social and political dimension to it and we expect member states to take different approaches on the detail of universal-service provisions. We have no problems with member states, or the market for that matter, going beyond minimum definitions of universal service. What matters in the context of costing and financing universal service is that there is a limit on what can be included in the calculation of the cost of universal-service obligations at a member-state level.

For the moment, the scope for this purpose will be the voice-telephony service and, of course, related elements such as the provision of emergency services, public payphones, directory enquiries, special schemes for needy users.

That does not mean it will not evolve over time. However, we are not yet near the stage where we can guarantee affordable switched broadband access to everyone. That requires enormous investments and no-one has yet proved it to be commercially viable in a mass market.

With respect to costing, the interconnection framework will set out the principles for calculating any cost associated with the imposition of universal-service obligations on an operator or operators at a national level. The basic principle will be to calculate the net cost imposed on the operator or operators with and without universal-service obligations, based on an avoidable-cost methodology.

With respect to the funding of universal service at a national level, the Commission believes that it may be possible for approaches to national funding schemes to vary from one member state to another, without creating significant barriers within the internal market for telecommunications infrastructure or services and without distorting competition.

Nonetheless, national regulatory authorities will be obliged to ensure publication of the details of the funding mechanisms being used, to notify them to the Commission in advance of their implementation, and to monitor them appropriately.

5.5. *Competition rules*

The fact that we are developing a regulatory framework for the 1998 liberalisation does not mean that there are no rules to deal with current business.

The competition rules, in particular Article 85 (agreements between undertakings) and Article 86 (abuse of dominant position) of the Treaty are applicable to all sectors. Mr Karel Van Miert has already made it very clear that he will apply the rules rigorously and that increased competition at a national level means that it is easier to justify alliances or mergers or to favourably interpret Article 86. In fact, proposed alliances such as Atlas are likely to be blocked until infrastructure competition occurs.

5.6. *A single Euro-regulator*

That leads on to the issue of a single Euro-regulator. It is true that the Bangemann Report called for an authority to be established at a European level.

In a way, we already have a European regulator. There is a considerable body of European telecommunications legislation and companies can go to national courts and the European Court to uphold it. In addition, the Treaty's Competition Rules and Merger Rules apply to the sector and case-work in the telecommunications sector is developing rapidly. Moreover, the ONP framework is also there to resolve conflicts of interest.

It may be that this issue will be dealt with in the 1996 Intergovernmental Conference, but any change will have to compatible with the Treaty and safeguard the independence of competition rules.

6. Lifting current constraints on the use of infrastructure in the provision of competitive telecommunications services

In advance of the political agreement to liberalise infrastructure, the Commission in its Autumn 1994 communication on lifting current constraints on the use of infrastructure in the provision of competitive telecommunications services (which subsequently became Part I of the infrastructure Green Paper) [6], proposed establishing a clear link between services liberalisation and infrastructure liberalisation. Such a link would imply that operators of service providers licensed or permitted to provide telecommunications services would have a free choice on the use of infrastructure to provide such services.

In practice, this would mean that cable-television networks or other telecommunications infrastructures could also be exploited to provide telecommunications services in the competitive area, ie, currently all services except (public) voice telephony.

It would also mean free infrastructure choice for providers of satellite and mobile communications as these are deemed to be competitive services as they lie outside the definition of telephony which is reserved (until 1998) in the 1990 services directive.

In fact, only part of the Commission's proposal was accepted at the November 1994 Council – the part linking full services liberalisation with full infrastructure liberalisation in 1998.

To properly examine the reason for the approach taken, it is necessary to examine both the background to the current proposal and the general logic of the approach to liberalisation of telecommunications to date. Only then is it possible to assess whether the political, legal, economic and social logic and objectives are mutually consistent (and desirable). Only then is it also possible to see how the approach may be extended to a comprehensive view of services and infrastructure competition (for example, including broadcasting, electronic or rather networked provision of information services generally), and examine whether the progressive approach coming from the telecommunications sectors is consistent with the plans and views of policy-makers in other parts of the so-called information sector.

The reader may also well ask whether the approach mirrors that in the United States where the combination of a number of fragmentary proposals would involve lifting the various constraints on entities crossing the borders between services and manufacturing, local and long distance, wireless and fixed, and telecommunications services and broadcasting services.

6.1. Telecommunications services and infrastructure liberalisation in the context of the European information society

A number of factors led to the Commission bringing forth its communication on infrastructure in the Autumn of 1994 which became Part I of the Green Paper.

Firstly, in the context of the resolution following the Telecommunications Review establishing the 1998 services deadline, four member states, France, Germany, the Netherlands and the UK requested that the Commission examine the use of cable television and alternative infrastructures for the delivery of telecommunications services already open to competition.

The Commission completed these studies in the Spring of 1994. The studies showed the varying approaches taken by member states in terms of permitting the exploitation of alternative infrastructures and the sometimes unclear rules governing which services cable-television networks are allowed to provide.

In general, the studies emphasised the contribution that these infrastructures could make to the telecommunications sector, in particular in putting downward pressure on the price of leased capacity, and in providing incentives to innovate and develop new telecommunications services (both to business and private consumers) in Europe.

The second background factor to Part I of the infrastructure Green Paper communication was the publication in 1994 of the Bangemann Report [12]. The report's first recommendation is to request member states to 'accelerate the on-going process of liberalisation of the Telecoms sector by opening up to competition infrastructures and services still in the monopoly area; removing non-commercial political burdens and budgetary constraints on telecommunications operators; and setting clear timetables and deadlines for the implementation of practical measures to achieve these goals'. The European Council in Corfu on the 24–25 June 1994 subsequently called for 'the necessary regulatory framework to be established at a Community level as soon as possible'.

In making its proposals in the infrastructure communication, the Commission wished to respond to the concerns about speeding up commercial development of the sector but without de-stabilising the political agreement on services liberalisation in 1998, which had become an adjustment target for operators. The Commission also wished to clarify proposals with respect to the use of cable-television networks and situate policy here within the overall framework, particularly given speculation that had arisen with respect to liberalisation in this area and given speculation concerning possible divestiture requirements on entities responsible both for telecommunications and cable-television networks.

6.2. *The particular case of cable-television networks*

During and subsequent to the Telecommunications Review in 1993 and given the development of cable-television networks offering telephony in the UK and speculation about the convergence of delivery networks, considerable pressure had built up, particularly from certain countries, for a Union-wide proposal on lifting constraints on cable-television providers. The situation in Europe is that levels of cable-television penetration are highly differentiated by country and that the ownership and control of networks also varies considerably.

Allowing cable-television networks to provide competitive telecommunications services obviously raised the question of whether at some point they would be allowed to provide public voice telephony and moreover, whether at some point, telecommunications operators would be permitted to provide broadcasting services. Hence the need for a progressive and consistent approach to lifting lines of business restrictions, which takes account of the commercial and technological reality of the situation and which is also consistent with the Treaty competition rules.

The development of Union policy and regulation in the telecommunications sector has been founded on a number of underlying principles. One is that regulation based on technological constraints which are likely to become redundant are to be avoided. A second is that where possible, there should be no line of business restriction and that specific competition concerns should be addressed under the competition rules.

An automatic requirement, for example, to divest cable television from telecommunications activities would, therefore, be counter to the general principles outlined. One apparent commercial and technological trend is a blurring of the distinction between telecommunications and cable-television networks. In the US there is the prospect of alliances between telecommunications operators and cable-television companies despite regulatory resistance to this phenomenon within a franchised area because of anti-trust concerns. Where competition issues arise (and they clearly do) there are alternatives to forced divestiture, such as requiring financial separation or allocating additional operational licenses in the markets concerned. Compulsory, rather than voluntary, divestiture is likely to lead to illogical consequences. For example, if a cable-television operator successfully extended its operations into the telephone area, would it automatically be obliged to divest itself?

On the other hand, the Commission is concerned, in the context of convergence, that member states may consider consolidating telecommunications and cable-television operations which have developed quite separately. Such a consolidation is likely to be inconsistent with the competition rules, unless additional infrastructure licenses are permitted, because it forecloses effective competition. One situation where consolidation could be permitted is between cable-television networks and parts of utility companies or other entities which have national networks to create a second nation-wide operator. Such a model appears to have been under consideration in the Netherlands and in Spain. But this does not mean that the Commission will favour any particular 'model', as the process of network convergence develops. Another perfectly rational consolidation or synergy model is between wireless operators and utility or cable-television companies or both.

The point of departure, therefore, for policy will be the competition rules together with the general logic and consistency of the progressive liberalisation of the sector, bearing in mind technological and market developments. Already, the casework in the telecommunications and related sectors is beginning to build-up. One thing is certain – there are only a limited number of entities which are in a position to finance the substantial investments required to operate widespread modern networks.

6.3. The logic and consistency of the progressive liberalisation of telecommunications

To best situate the current proposals on infrastructure liberalisation and the legal reforms that are implied, it is important to trace the progressive approach to regulatory change.

The major liberalising steps have occurred by means of Article 90 directives in 1988 (equipment) and 1990 (services) and both these directives have been upheld by the European Court with the limited qualifications that there was a need to define special (as opposed to exclusive) rights and that contractual arrangements should be dealt with under Articles 85 and 86.

The 1990 services directive had a stepwise aspect to it with the possibility of delaying resale of capacity for data services until 1993, to allow adjustment between leased-circuit tariffs and tariffs for data service. From the beginning of 1993, the only services area that could be excluded from competition was (public) voice telephony as defined, and following the Telecommunications Review this exclusion will also cease in 1998. Apart from the fact that telex service was exempted from the provisions of the services directive, five major exclusions were made in the 1990 directive:

1) Voice telephony as defined in the directive was excluded from the withdrawal of special or exclusive rights (Article 2); this exclusion will now cease in 1998.
2) Mobile radiotelephony and paging was excluded (Article 1 (2)).
3) Satellite services were excluded (Article 1 (2)). This exclusion ceased following the adoption on the 13 October 1994 of the Article 90 directive on satellite communications [13].
4) Radio-broadcasting and television were excluded from the definition of telecommunications services (Article 1 (1)).
5) The directive focused only on the provision and interconnection of leased lines by the telecommunications organisations granted special or exclusive rights for the provision of public telecommunications networks (ie, public telecommunications infrastructure) (Article 4).

The satellite Green Paper [14] and subsequent resolution extended the provisions of the services directive into the satellite field. In legal terms, equipment and non-voice telephony services in the satellite area were liberalised at the end of 1994 by means of an Article 90 directive modifying the 1990 services directive.

In addition, in November 1994 the Commission adopted in first reading a proposal to further amend the 1990 services directive to abolish the restrictions on the use of cable-television networks to provide telecommunications services.

The Commission has also continued to argue strongly that GSM services should be subject to competitive provision and in fact a number of member states have been instructed to award competitive licences. In addition, the 1994 mobile Green Paper has proposed an overall framework of policy for all mobile and personal communications. The Commission has now come forward with a formal proposal to modify the 1990 services directive to allow free choice on infrastructure for mobile communications, implicitly lifting exclusion 2) above and removing part of exclusion 5).

Part I of the infrastructure Green Paper on lifting constraints on the use of infrastructure is directed essentially at lifting the whole of the fifth exclusion listed above. On 19 July 1995, the Commission adopted in first reading the proposed Article 90 directive on full liberalisation. This covers both the lifting of the first exclusion in 1998 and the full lifting of the fifth exclusion in 1996.

The proposal for linking the lifting of constraints on infrastructure with currently liberalised services was intended to apply not only to mobile communications and to cable-television networks but to the use of alternative infrastructure such as transport and utility networks to provide non-voice telephony.

When the first or voice telephony exclusion is lifted in 1998, current proposals would imply that only the fourth exclusion (the exclusion of broadcasting) remains in place.

6.4. *How robust is the logic and what next?*

So much for the context of the Autumn 1994 proposal that became the first part of the infrastructure Green Paper. The logic is one that is built on the services directive (and Treaty rules) and is based on rolling back exclusive rights and practices where they have been extended to new service areas or lifting constraints where technological change makes restrictions inappropriate or limits innovation, providing that the lifting of constraints does not endanger the financial ability of entities required to provide public-service provisions. To that extent it has both legal and economic logic.

Although there was political agreement on full liberalisation in 1998, no consensus emerged on the other part of the Commission's proposal to fully lift exclusion 5) immediately so that other infrastructures could be used for liberalised service areas. The Commission has not however given up on its wish to accelerate part of the infrastructure liberalisation. This is the reason for the proposal on cable television, the one on mobile and the subsequent one on full liberalisation. In legal and political terms, it is more difficult to accelerate the process for the other alternative infrastructures such as utilities, etc. However, if nothing happens here, there is the clear danger that the 1998 liberalisation will not be immediately effective because it will take time for other infrastructure-based telecommunications operators to emerge.

The legal logic of the whole exercise means that exclusion 1) will be lifted from the beginning of 1998. In addition, so will exclusion 5) if it has not been fully lifted before that date. Exclusion 2) on mobile should be lifted in 1996, although it may be that the prohibition on mobile operators to directly connect their networks between member states will not be fully lifted until the 1998 deadline.

A reasonably predictable liberalisation process is under way, therefore, that will see all the restrictions lifted except for exclusion 4) on broadcasting. With the gradual emergence of multimedia services, this line of business restriction is also now on the agenda. It was dealt with in a very sketchy fashion in Part II of the infrastructure Green Paper and as Figure 1 shows it still constitutes a clearly separate activity apart from the pending entry of some cable-television operators into telephony. However, the issue of allowing entities to cross the divide is at the heart of the discussions on reform of the Communications Act in the USA. And the matter continues to be discussed in the UK despite the insistence of the government that the timed exclusion of British Telecommunications from providing broadcast services

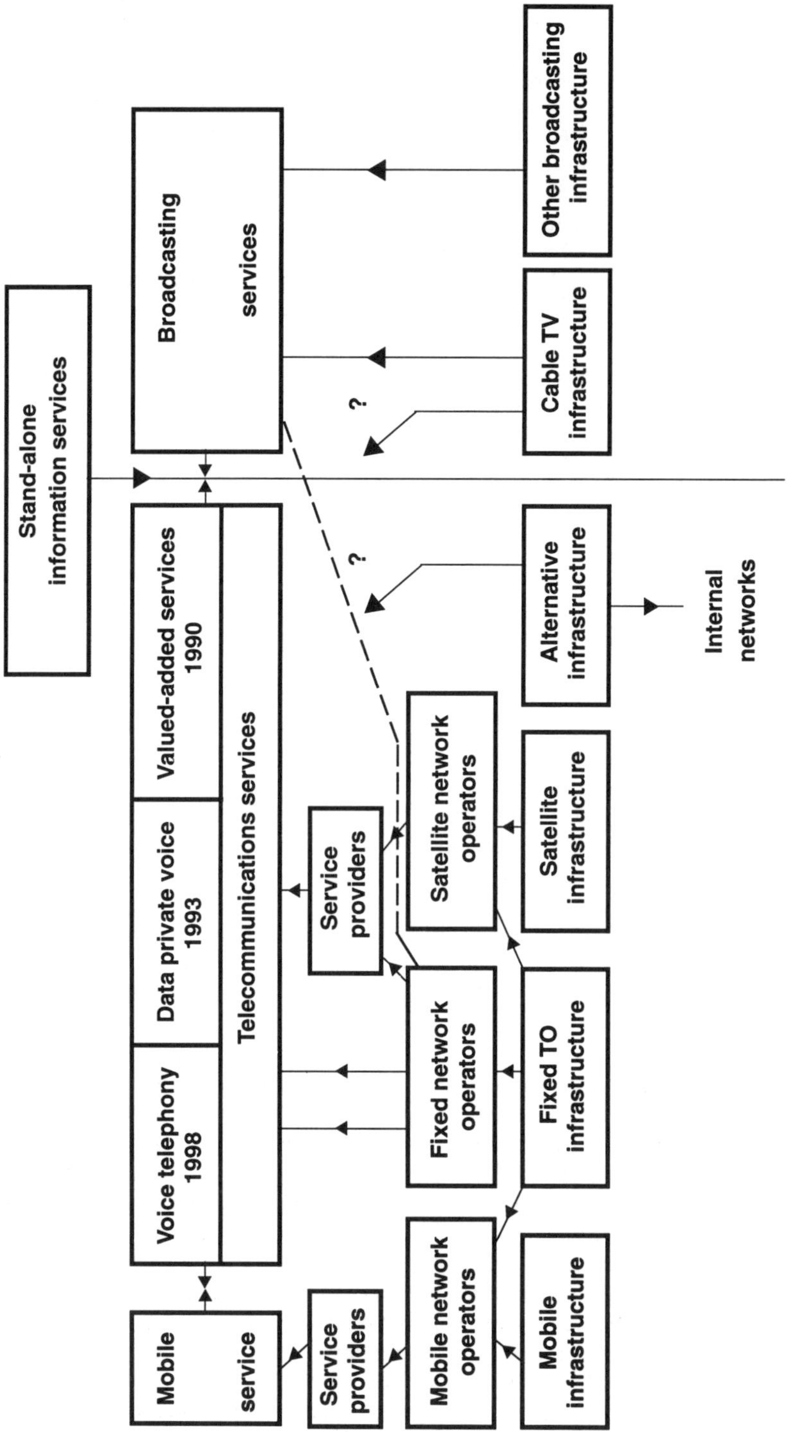

Figure 1. Lifting regulatory restrictions

over its network established during the 1991 review will not be reconsidered.

Two major kinds of convergence are taking place. Firstly, the world of personal communications implies that classic distinctions between fixed and wireless networks or fixed and wireless-service providers are becoming increasingly redundant. The second is that the broadening of telecommunications-service provision first to data and value-added services and then to visual signals means that it will become harder to distinguish between telecommunications 'one-to-one, two-way' networks and broadcasting or cable-television 'one-to-many, one-way' networks.

In addition to telecommunications and broadcasting networks, there are also substantial parts of the information sector which could be currently described as 'stand-alone' services. These include large parts of publishing, entertainment and communications which are supplied on a physical basis without the use of electronic networks. However, the increasing digitalisation of information and the advances in cost-effective computing power made by the PC and related electronic devices means that not only are new information services being created but that parts of the existing information sector may in the future be provided in electronic form and indeed provided or transmitted in networked electronic form. An example of the first is a book on CD-ROM, an example of the second is the distribution of news publications over networks or indeed postal material transmitted for part of its journey by electronic rather than physical means.

Figure 1 is therefore illustrating two types of key questions. The first is the one that is rightly at the centre of the infrastructure Green Paper, ie, what safeguards are required in the lifting of constraints on the provision of infrastructure for all telecommunications services, including voice telephony?

By 'safeguards' is meant both the means of ensuring that the financial viability of telecommunications organisations entrusted with particular social obligations is not undermined, and that there are means to promote effective competition whilst paying attention to resource constraints or issues of public or consumer interest.

The main areas to be addressed have been outlined in the infrastructure communication. They include (a) a framework for interconnection, (b) rules of the game for sharing universal-service obligations, (c) a framework for licensing issues, and (d) open and comparable access to third markets.

The second question is the one related to the last remaining exclusion in the services directive and to the prospective convergence between telecommunications and other parts of the information sector including broadcasting, ie, how to facilitate the development of integrated communications services and the networks that are required to carry them.

For the moment, the main constraint on the development of some of these new services is excessively-priced transmission capacity and access which in turn is the result of limited infrastructure competition. In addition, there are a number of related questions, such as intellectual property rights, information security, data protection and privacy issues and, of course, broader competition issues including cross-ownership.

Although telecommunications operators are free to develop new integrated or multimedia or information services over their own networks, the fact that market and service developments are linked very closely to digitalisation and related technologies and the fact that there are obvious economic synergies with digital television explains why the debate is also moving towards the future regulation of broadcasting. In any case, for all the talk of multimedia services the only services that generate large revenues at the moment are voice telecommunications and (cabled) television services.

Ultimately, the question will be what physical and spectrum resources are allocated to the provision of a wide range of services which embraces the traditional telecommunications and broadcasting services as well as all the new ones.

This second question pertaining to broadcasting has been briefly considered in the infrastructure Green Paper but it is quite separate from the first one dealing with regulation linked to infrastructure liberalisation. More specifically, it moves into uncharted territory for the Commission and emphasises the important distinction between conveyance and content. The question can also be separated in another fashion. The liberalisation step implied by the first question is opposed by many TOs and supported by cable-television operators. The step implied by the second question would be supported by TOs, opposed by many cable-television operators and raise concerns amongst many broadcasters. However, it is only by dealing with the second one as well as the first one before the end of the decade that Europe can expect to keep abreast of North America in its search for a more open and consistent regulatory approach to the whole of the communications sector.

7. Additional considerations

One major omission of this chapter has been a consideration of the additional, more detailed measures needed to accompany and make effective the progressive lifting of regulatory constraints, be they on service provision or infrastructure provision. Some have been mentioned – dealing with interconnection and maintaining and financing universal service. Others such as reciprocity and trade issues have been omitted.

To some extent there is a parallel with the telecommunications-equipment sector. The declaration that all would be liberalised was the easy part. The actual detail of moving towards a liberalised European or single market, including the mutual recognition of testing and type approval and the development of common or mutually-recognised standards is far more complex.

In the services and networks domain, the additional and accompanying measures concern, amongst others, standards, frequency reservation, numbering and licensing. These issues have a national dimension but a wider European one as well. They form a substantial part of regulatory oversight and merit consideration in their own right. The same applies to another set of issues which arise in all economic sectors but which are becoming increasingly pertinent in the communications sector, ie, content, intellectual property rights, privacy, data and consumer protection.

On the latter set of issues, policy at a European level is also being tackled by other parts of the Commission in particular DGs III, X and XV – for an overview, see the action plan for the information society [15]. In this context, a Green Paper on pluralism and media concentration in the international market was published in December 1992 [16] and a Green Paper on strategy options to strengthen the European programme industry in the context of the audiovisual policy of the European Union was published in April 1994 [17]. Many will have followed the internal wranglings within the Commission concerning the appropriate policy in the audiovisual area which has focused on quotas in the broadcasting area but which really is about getting the right balance between the free movement of services and industrial-policy concerns. In March 1995, the Commission adopted an audiovisual policy-package which combined direct support for industry with proposed modifications to the television without frontiers directive [18]. In addition, European-level legislation is continuing to develop in areas such as copyright, intellectual property rights, consumer and data protection. In July 1995, the Commission adopted its Green Paper on the future of copyright in information services. More will follow.

8. Conclusions

The major conclusion must be that with respect to pricing or tariffing, the work focus of the Commission is now firmly on putting in place the necessary legal and regulatory framework to ensure that there is effective competition in the telecommunications sector.

In that sense, the priorities have shifted towards the three major issues of interconnection, licensing and universal service. The work programme is concentrating on putting the necessary framework in place. This includes necessary recourse procedures and, of course, full back-up by competition rules.

The emphasis will be on minimum regulation with maximum clarity whilst recognising that in a transitional period to competition, substantial regulatory oversight of the dominant incumbent operators is important.

The outstanding issue to be resolved before the end of 1995 is to what extent the liberalisation process can be accelerated by having a two-step infrastructure liberalisation with immediate liberalisation (effectively 1996) of mobile and alternative infrastructures.

In addition, the Commission will now begin to look at the potential regulatory problems linked to the development of multimedia services. Many telecommunications regulators are also beginning to do the same.

On pricing issues per se, the Commission will essentially intervene on the pricing of bottleneck resources or where there is substantial market power by incumbent operators who risk stalling the introduction of effective competition. The Commission in this sense will look out for over-pricing, under-pricing or predatory-pricing and unreasonable cross-subsidisation. So the emphasis will be on the pricing of leased capacity and interconnection.

However much the Commission dislikes tariffing levels and structures in some retail areas, and international and cross-border tariffs in Europe are an obvious example, it is unlikely to interfere in this domain but instead work towards the introduction of effective competition. This includes competition in the mobile area and encouraging direct interconnection of competing national networks, and it also includes the substitution from 1998 of interconnection arrangements as opposed to the traditional accounting-rate agreements between member states.

Notes and references

[1] Proposed amendment of Directive 90/388/EEC concerning mobile and personal communications, adopted by the Commission 21 June 1995.

[2] Proposed amendment of Directive 90/388/EEC abolishing restrictions on the use of cable-television networks to provide telecommunications services, *Official Journal of the European Communities* C76/8, 28 March 1995.

[3] Proposed amendment of Directive 90/388/EEC regarding the implementation of full competition in telecommunications markets.

[4] Communications Outlook 1995. OECD, Paris, 1995.

[5] Towards the personal communications environment: Green Paper on a common approach to mobile and personal communications in the European Union, COM(94) 145, 27 April 1994.

[6] Communication to the Council and European Parliament on the liberalisation of telecommunications infrastructure and cable-television networks, Part I Green Paper, COM(94) 440, 25 October 1994.

[7] Communication to the Council and European Parliament on the liberalisation of telecommunications infrastructure and cable-television networks, Part II Green Paper, COM(94) 682, 25 January 1995.

[8] Communication on the consultation on the Green Paper on the liberalisation of telecommunications infrastructure and cable-television networks, COM(95) 158, 3 May 1995.

[9] Commission Directive of 28 June 1990 on competition in the markets for telecommunications services, 90/388/EEC, *Official Journal of the European Communities* L192/10, 24 July 1990.

[10] Council Directive of 5 June 1992 on the application of open network provision to leased lines, 92/44/EEC, *Official Journal of the European Communities* L165/27, 19 June 1992.

[11] Proposed Directive on interconnection in the context of ONP, COM(95) 379, adopted by the Commission 19 July 1995.
[12] Europe and the global information society: Recommendations to the European Council (the Bangemann Report), Brussels, 26 May 1994.
[13] Commission Directive 94/46/EEC amending Directive 88/301/EEC and Directive 90/388/EEC in particular with regard to satellite communications, 13 October 1994.
[14] Towards Europe-wide systems and services: Green Paper on a common approach in the field of satellite communications in the European Community, COM(90) 490, 20 November 1990.
[15] Europe's Way to the Information Society, an Action Plan: Communication by the European Commission, COM(94) 347, 19 July 1994.
[16] Pluralism and media concentration in the international market: Commission Green Paper, COM(92) 480, 23 December 1992.
[17] Strategy options to strengthen the European programme industry in the context of the audiovisual policy of the European Union: Commission Green Paper, 7 April 1994.
[18] Communication on the revision of Directive 89/552/EEC (television without frontiers) and Communication on a regulatory framework for services for the information society, COM(94) 524 and COM(95) 444, 22 March 1995.

List of contributors

David Allen	Concord, Massachusetts, USA
Jacques Arlandis	IDATE, Montpellier, France
Jean-Claude Burgelman	Studies on Media, Information and Telecommunication, Free University of Brussels, Belgium
Richard A. Cawley	European Commission, Brussels, Belgium
Andrea Costa	Università Cattolica del Sacro Cuore, Milan, Italy
Krisztina Heller	Hungarian Telecommunications Company, Budapest, Hungary
Bridger Mitchell	Charles River Associates, Palo Alto, California, USA
Franco Morganti	Databank Consulting, Milan, Italy
Werner Neu	Wissenschaftliches Institut für Kommunikationsdienste, Bad Honnef, Germany
Karl-Heinz Neumann	Wissenschaftliches Institut für Kommunikationsdienste, Bad Honnef, Germany
Laura Solimene	Università Cattolica del Sacro Cuore, Milan, Italy
Pascal Verhoest	Studies on Media, Information and Telecommunication, Free University of Brussels, Belgium
Ingo Vogelsang	Boston University, Boston, USA

Author Index